Flask开发
Web搜索引擎
入门与实战

张子宪 编著

清华大学出版社
北京

内容简介

本书介绍如何学习和使用流行的 Flask 框架开发搜索引擎应用，主要内容包括面向 Web 开发的 Python 编程语言入门，使用 Python 构建 REST API，使用 Flask-RESTPlus 生成 Swagger 文档，搜索引擎应用前端展示实现及自动完成功能与拼写纠错技术，互联网搜索引擎案例分析。

全书共分 7 章：第 1 章着重介绍如何使用 Python 和 Elasticsearch 开发搜索引擎应用；第 2 章着重介绍 Python 的基本语法及其在 Web 开发中的使用；第 3 章着重介绍 Web 应用程序框架 Flask 和模板引擎 Jinja，以及构建 REST API 方法；第 4 章着重介绍 Werkzeug 库和 Flask 框架的源代码；第 5 章着重介绍 SQLAlchemy 和 Flask-SQLAlchemy 扩展；第 6 章着重介绍 Elasticsearch 的前端展示实现及自动完成功能与拼写纠错技术；第 7 章着重介绍医药垂直搜索引擎和集成了 Elasticsearch 的内容管理系统搜索 CastleCMS。

本书适合需要具体实现搜索引擎应用的开发人员或者对人工智能等相关领域感兴趣的人士参考。

本书封面贴有清华大学出版社防伪标签，无标签者不得销售。
版权所有，侵权必究。举报：010-62782989，beiqinquan@tup.tsinghua.edu.cn。

图书在版编目（CIP）数据

Flask 开发 Web 搜索引擎入门与实战 / 张子宪编著. —北京：清华大学出版社，2022.4
ISBN 978-7-302-60132-6

Ⅰ. ①F… Ⅱ. ①张… Ⅲ. ①软件工具－程序设计 Ⅳ. ①TP311.561

中国版本图书馆 CIP 数据核字（2022）第 025881 号

责任编辑：张　敏
封面设计：吴海燕
责任校对：徐俊伟
责任印制：朱雨萌

出版发行：清华大学出版社
网　　址：http://www.tup.com.cn, http://www.wqbook.com
地　　址：北京清华大学学研大厦 A 座　　邮　编：100084
社　总　机：010-83470000　　邮　购：010-62786544
投稿与读者服务：010-62776969, c-service@tup.tsinghua.edu.cn
质 量 反 馈：010-62772015, zhiliang@tup.tsinghua.edu.cn
印 装 者：艺通印刷（天津）有限公司
经　　销：全国新华书店
开　　本：185mm×260mm　　印　张：9.75　　字　数：216 千字
版　　次：2022 年 6 月第 1 版　　印　次：2022 年 6 月第 1 次印刷
定　　价：59.80 元

产品编号：089799-01

前言

  Flask 是一个使用 Python 编程语言实现的轻量级 Web 应用程序框架。使用 Flask 开发 Web 搜索引擎是一种可能的选择。本书介绍了 Flask 结合 Elasticsearch 搜索服务器开发搜索引擎应用。

  本书共分 7 章：第 1 章介绍如何使用 Python 和 Elasticsearch 开发搜索引擎应用；第 2 章介绍 Python 基本语法及其在 Web 开发中的使用；第 3 章介绍 Web 应用程序框架 Flask 和模板引擎 Jinja，以及使用 Flask 构建 REST API 和使用 Flask-RESTPlus 生成 Swagger 文档；第 4 章分析 Werkzeug 库和 Flask 框架的源代码；第 5 章介绍 SQLAlchemy 和 Flask-SQLAlchemy 扩展；第 6 章介绍 Elasticsearch 的前端展示实现及自动完成功能与拼写纠错技术；第 7 章介绍医药垂直搜索引擎和集成了 Elasticsearch 的内容管理系统 CastleCMS。

  由于作者水平有限，书中难免有疏漏之处，敬请广大读者谅解。

  感谢早期合著者、合作伙伴、员工、学员、读者的支持，给我们提供了良好的工作基础。就像玻璃容器中的水培植物一样，这是一个持久可用的成长基础。技术的融合与创新无止境。欢迎一起探索。

编者

目录

第1章 Web 搜索引擎开发 ·· 1
  1.1 准备工作环境 ·· 1
  1.2 Linux 操作系统基础 ·· 2
  1.3 Elasticsearch 的 Python 客户端 ·· 3
    1.3.1 安装 Elasticsearch ·· 3
    1.3.2 基本使用 ··· 5
    1.3.3 定义索引结构 ··· 12

第2章 Python 技术基础 ·· 16
  2.1 变量 ·· 16
  2.2 注释 ·· 16
  2.3 简单数据类型 ·· 16
    2.3.1 数值 ·· 17
    2.3.2 字符串 ·· 19
    2.3.3 数组 ·· 21
  2.4 字面值 ·· 21
  2.5 控制流 ·· 22
    2.5.1 if 语句 ·· 22
    2.5.2 循环 ·· 23
  2.6 列表 ·· 24
  2.7 元组 ·· 25
  2.8 集合 ·· 26
  2.9 字典 ·· 27
  2.10 位数组 ··· 29
  2.11 模块 ·· 30
  2.12 函数 ·· 30
    2.12.1 print 函数 ·· 31
    2.12.2 定义函数 ··· 32

2.13 面向对象编程 ··· 34
　　2.13.1 静态方法 ··· 36
　　2.13.2 __call__方法 ··· 37
2.14 使用 StringIO 模块 ·· 38
2.15 文件操作 ··· 39
　　2.15.1 读写文件 ··· 39
　　2.15.2 重命名文件 ··· 41
　　2.15.3 遍历文件 ··· 41
2.16 迭代器 ·· 42
　　2.16.1 zip函数 ·· 43
　　2.16.2 itertools模块 ··· 44
2.17 数据库 ·· 45
2.18 日志 ·· 48

第 3 章 Flask 框架与微服务 50
3.1 Flask 简介 ·· 50
3.2 模板引擎 Jinja ··· 51
　　3.2.1 Jinja的基本使用 ··· 52
　　3.2.2 实现分页 ·· 54
　　3.2.3 在Flask中使用Jinja ··· 56
3.3 测试 RESTful API 的 curl 命令 ······································· 56
3.4 JSON 数据格式 ·· 58
3.5 构建 REST API ·· 59
3.6 Swagger 文档 ·· 62
3.7 使用 Fetch API ·· 72
3.8 发布 Flask 到 Nginx ··· 74
3.9 启用 HTTPS ··· 74

第 4 章 Flask 源代码分析 76
4.1 Werkzeug 库 ·· 76
　　4.1.1 WSGI简介 ··· 77
　　4.1.2 Werkzeug演示 ·· 77
4.2 源代码分析 ··· 81

第 5 章 SQLAlchemy 操作数据库 85
5.1 使用 SQLAlchemy ··· 85

5.2	SQL 表达式语言	85
	5.2.1 定义和创建表	86
	5.2.2 模式	87
	5.2.3 插入和查询	88
5.3	Flask-SQLAlchemy 扩展	90

第 6 章 Elasticsearch 分布式搜索引擎 ... 91

6.1	实现用户界面	91
	6.1.1 搭建JavaScript环境	91
	6.1.2 Node.js基础	97
	6.1.3 使用React前端库	101
	6.1.4 使用webpack模块捆绑器	124
6.2	自动完成	125
6.3	拼写纠错	132
	6.3.1 模糊匹配问题	132
	6.3.2 英文拼写检查	142
	6.3.3 中文拼写检查	143

第 7 章 Web 搜索案例分析 ... 145

7.1	医药垂直搜索引擎	145
7.2	内容管理系统搜索	147

第 1 章
Web 搜索引擎开发

Elasticsearch 是一个分布式和高可用的搜索引擎。本章介绍如何使用 Python 语言和 Elasticsearch 实现 Web 搜索引擎。

1.1　准备工作环境

首先要准备一个 Python 的开发环境。当前可以使用 Python 3.9 版本。Python 3.9 可以从 Python 的官方网站 https://www.python.org/下载得到。使用默认方式安装即可。

在 Windows 下安装 Python 以后，在控制台输入 python 命令进入交互式环境。

```
C:\Users\Administrator>python
Python 3.9.6 (tags/v3.9.6:db3ff76, Jun 28 2021, 15:26:21) [MSC v.1929 64 bit (AMD64)] on win32
Type "help", "copyright", "credits" or "license" for more information.
>>>
```

在 Windows 操作系统下，利用如下命令检查 Python 3 是否已经正确安装，及其版本号：

```
>python3 -V
Python 3.9.6
```

检查 Python 3 所在的路径：

```
>where python3
C:\Users\Administrator\AppData\Local\Microsoft\WindowsApps\python3.exe
```

可以准备一个用于编写代码的集成开发环境。例如，可以使用 PyCharm 或者 Visual Studio。也可以使用 Notepad++这样的文本编辑器写 Python 代码。

1.2 Linux 操作系统基础

很多 Web 应用的后台运行在 Linux 操作系统中。Linux 来源于 UNIX，是 UNIX 操作系统的开放源代码实现。Linux 通过 SSH 客户端软件连接到远程的 Linux 服务器。SSH 服务器通常作为大多数 Linux 发行版上易于安装的软件包提供。可以尝试使用 ssh localhost 命令来测试它是否正在运行。

如果有现成的 Linux 服务器可用，可以使用支持 SSH（Secure Shell，安全外壳）协议的终端仿真程序 SecureCRT 连接到远程 Linux 服务器。因为它可以保存登录密码，所以使用比较方便。除了 SecureCRT，还可以使用开源软件 PuTTY（http://www.chiark.greenend.org.uk/~sgtatham/putty），或者保存登录密码的 KiTTY（https://www.fosshub.com/KiTTY.html）及 Xshell。如果用 root 账户登录，则终端提示符是#；否则，终端提示符是$。

也可以在 Windows 下安装 Cygwin，使用它来练习 Linux 常用命令。

小袋鼠在袋鼠妈妈的袋子里长大。使用 VMware，Linux 可以运行在 Windows 系统下。VMware 让 Linux 运行在虚拟机中，而且不会破坏原来的 Windows 操作系统。首先要准备好 VMware，当然仍然需要 Linux 光盘文件。

就好像华山派有剑宗和气宗，Linux 也有很多种版本，如 RedHat、CentOS、Ubuntu 及 SUSE。这里介绍 Ubuntu（https://www.ubuntu.com）和 CentOS（http://www.centos.org/）。

操作系统中可能会安装好几个版本的 JDK。在 Linux 中，为了切换 JDK 版本，只需要修改/etc/alternatives 中的符号链接指向。

在 Ubuntu 中，如果需要安装软件，可以下载 deb 安装包，然后使用 dpkg 命令安装。但一个软件包可能依赖其他的软件包。为了安装一个软件可能需要下载其他的好几个它所依赖的软件包。

为了简化安装操作，可以使用高级包装工具（Advanced Packaging Tool，APT）。APT 会自动计算出程序之间的相互关联性，并且计算出完成软件包的安装需要哪些步骤。这样在安装软件时，不会再被那些关联性问题所困扰。

在/etc/apt/sources.list 文件中指示了包的来源存储库。包的来源可以是 CD 或 DVD、硬盘上的目录、HTTP 或 FTP 服务器上的目录。请求的数据包位于服务器或本地硬盘上，它将自动下载并安装。APT 主要关注采购包、包的可用版本的比较，以及包档案的管理。实际上，可以通过浏览器浏览在 HTTP 或 FTP 服务器上的存储库。

如果需要修改/etc/apt/sources.list 文件，可以先备份这个文件：

```
#sudo cp /etc/apt/sources.list /etc/apt/sources.list.bak
```

如果这一步出现

```
sudo: unable to resolve host t-000004
```

这样的错误，则可以考虑执行如下命令修改/etc/hosts 文件的内容：

```
#echo $(hostname -I | cut -d\  -f1) $(hostname) | sudo tee -a /etc/hosts
```

如果安装过程中出现 "E: Could not get lock /var/lib/dpkg/lock" 这样的错误，则可以尝试使用如下命令修复：

```
#sudo fuser -cuk /var/lib/dpkg/lock
#sudo rm -f /var/lib/dpkg/lock
```

在 CentOS 中，如果需要安装软件，可以下载 RPM 安装包，然后使用 RPM 安装。例如，下载 Elasticsearch 软件的安装包 elasticsearch-6.6.0.rpm：

```
#wget https://artifacts.elastic.co/downloads/elasticsearch/elasticsearch-6.6.0.rpm
```

使用如下命令安装：

```
#rpm -ivh elasticsearch-6.6.0.rpm
```

但有些操作系统对应的 RPM 安装包找起来比较麻烦。一个软件包可能依赖其他的软件包。为了安装一个软件可能需要下载其他的好几个它所依赖的软件包。

为了简化安装操作，可以使用黄狗升级管理器（Yellow dog Updater，Modified），一般简称 yum。yum 会自动计算出程序之间的相互关联性，并且计算出完成软件包的安装需要哪些步骤。这样在安装软件时，不会再被那些关联性问题所困扰。

yum 软件包管理器自动从网络下载并安装软件。yum 有点类似 360 软件管家，但是不会有商业倾向的推销软件。例如，安装支持 wget 命令的软件：

```
#yum install wget
```

为了方便在服务器端编写 Shell 脚本，可以采用 Micro（https://github.com/zyedidia/micro）这样的终端文本编辑器。

在 Linux 上，可以通过 snap 安装 micro：

```
#snap install micro --classic
```

保存文件后，按 Ctrl+Q 组合键退出。

1.3 Elasticsearch 的 Python 客户端

首先介绍 Elasticsearch 的安装以及 Python 客户端的基本使用，然后介绍如何定义索引结构。

1.3.1 安装 Elasticsearch

如果在 Windows 操作系统下，则可以从 https://www.elastic.co/cn/downloads/elasticsearch 下载 Elasticsearch 的安装包。这里使用的版本为 7.8.0。得到文件 elasticsearch-7.8.0.zip。

直接解压至某目录，例如 D:\elasticsearch-7.8.0。下载完解压后有以下几个路径：bin 是运行的脚本；config 是设置文件；lib 中放的是依赖包。到目录 D:\elasticsearch-7.8.0\bin 下，运行 elasticsearch.bat。

如果显示 Java 虚拟机（Java Virtual Machine，JVM）内存不够，则可以在 D:\elasticsearch-7.8.0\config\jvm.options 配置文件中调整内存大小。其中，Xms 参数表示堆空间的初始大小；Xmx 参数表示堆空间的最大大小。应该把最小和最大 JVM 堆设置成相同的值。例如：

```
-Xms4g
-Xmx4g
```

成功启动 Elasticsearch 后，在浏览器中打开网址：http://localhost:9200/。

启动成功后，会在解压目录下增加 2 个文件夹：data 用于存储索引数据；logs 用于日志记录。因为创建索引耗时，所以预先将文档写入一个日志目录。

如果使用 Linux 的 Ubuntu 发行版，则可以用 debian 安装包安装 Elasticsearch；如果使用 RedHat Enterprise Linux，则可以使用 RPM 包安装 Elasticsearch。

在 Windows 操作系统下，启动 Elasticsearch 服务：

```
> bin\elasticsearch.bat
```

通过 HTTP（Hyper Text Transfer Protocol，超文本传输协议）发送指令和 Elasticsearch 交互。curl 是一个知名的网络命令行工具，可以用来发送 GET 或者 POST 命令。在 Linux 下默认已经安装了这个命令行工具，可以用 curl 命令行访问网址 http://localhost:9200/：

```
> curl http://localhost:9200/
```

或者在命令行外壳程序 PowerShell 中运行：

```
> Invoke-RestMethod http://localhost:9200
```

在 Windows 下可以用 Python 代码测试 Elasticsearch 服务：

```
import requests
url = 'http://localhost:9200'

response = requests.get(url)

print(response.status_code)
print(response.content)
```

Kibana 是 Elasticsearch 的开源数据可视化仪表板，可以在 Kibana 控制台发送命令给 Elasticsearch。在 Windows 操作系统下，Kibana 可以使用 .zip 软件包安装，从 https://www.elastic.co/cn/downloads/kibana 下载 Kibana 安装包。

在编辑器中打开配置文件 config/kibana.yml 设置 Elasticsearch 实例：

```
elasticsearch.hosts: ["http://localhost:9200"]
```

在 Windows 操作系统下运行 bin\kibana.bat，以启动 Kibana。

在浏览器中打开网址 http://localhost:5601 进入 Kibana 的管理界面。

如果无法正常访问，则可以用如下命令检查端口占用情况。

```
>netstat -ano | findstr 5601
```

如果仍然无法正常使用，则可以先测试 X-Pack 插件或者检查 Elasticsearch 中插件的安装情况。

1.3.2 基本使用

elasticsearch-py（https://github.com/elastic/elasticsearch-py）是 Elasticsearch 的官方底层 Python 客户端。Elasticsearch DSL（https://github.com/elastic/elasticsearch-dsl-py）是高层客户端。

先介绍 elasticsearch-py 的使用。使用 pip 命令安装 elasticsearch 模块：

```
pip install elasticsearch
```

定义索引：

```
from elasticsearch import Elasticsearch

#By default we connect to localhost:9200
es = Elasticsearch()

es.indices.create(index='accesslog', ignore=400)

es.indices.put_mapping(
    index="accesslog",
    body={
        "properties": {
            "logdate": {
                "type":"date",
                "format":"dd/MM/yyy HH:mm:ss"
            }
        }
    }
)
#删除索引
es.indices.delete(index='accesslog')
```

简单的用法如下：

```
>>> from datetime import datetime
>>> from elasticsearch import Elasticsearch

#默认情况下连接到localhost:9200
>>> es = Elasticsearch()
```

```
#在elasticsearch中创建一个索引,忽略状态码400(索引已经存在)
>>> es.indices.create(index='my-index', ignore=400)
{u'acknowledged': True}

#会序列化datetime
>>> es.index(index="my-index", doc_type="test-type", id=42, body={"any": "data", "timestamp": datetime.now()})
{u'_id': u'42', u'_index': u'my-index', u'_type': u'test-type', u'_version': 1, u'ok': True}

#但没有反序列化
>>> es.get(index="my-index", doc_type="test-type", id=42)['_source']
{u'any': u'data', u'timestamp': u'2016-05-12T19:45:31.804229'}
```

索引和查询员工的例子:

```
from elasticsearch import Elasticsearch

es = Elasticsearch()

e1={
    "first_name":"nitin",
    "last_name":"panwar",
    "age": 27,
    "about": "Love to play cricket",
    "interests": ['sports','music'],
}
print(e1)

#插入文档
res = es.index(index='megacorp',id=1,body=e1)
print(res)

e2={
    "first_name" :  "Jane",
    "last_name" :   "Smith",
    "age" :         32,
    "about" :       "I like to collect rock albums",
    "interests":  [ "music" ]
}

res=es.index(index='megacorp',id=2,body=e2)
print(res)

e3={
    "first_name" :  "Douglas",
```

```
        "last_name" :   "Fir",
        "age" :         35,
        "about":        "I like to build cabinets",
        "interests": [ "forestry" ]
}
res=es.index(index='megacorp',id=3,body=e3)

res=es.get(index='megacorp',id=3)
print(res)

#检索文档
res= es.search(index='megacorp',body={'query':{'match_all':{}}})
print("Got %d Hits:" % res['hits']['total']['value'])

res= es.search(index='megacorp',body={'query':{'match':{'first_name':'nitin'}}})
print(res['hits']['hits'])

#布尔运算符
res= es.search(index='megacorp',body={
    'query':{
        'bool':{
            'must':[{
                'match':{
                    'first_name':'nitin'
                }
            }]
        }
    }
})

print(res['hits']['hits'])

#过滤运算符
res= es.search(index='megacorp',body={
    'query':{
        'bool':{
            'must':{
                'match':{
                    'first_name':'nitin'
                }
            },
            "filter":{
                "range":{
                    "age":{
                        "gt":25
                    }
```

```python
                }
            }
        }
    })
print(res['hits']['hits'])

#增加数据,尝试全文搜索
e4={
    "first_name":"asd",
    "last_name":"pafdfd",
    "age": 27,
    "about": "Love to play football",
    "interests": ['sports','music'],
}
res=es.index(index='megacorp',id=4,body=e4)
print(res)

res= es.search(index='megacorp',body={
    'query':{
        'match':{
            "about":"play cricket"
        }
    }
})
for hit in res['hits']['hits']:
    print(hit['_source']['about'] )
    print(hit['_score'])
    print('*********************')

#短语匹配
res= es.search(index='megacorp',body={
    'query':{
        'match_phrase':{
            "about":"play cricket"
        }
    }
})
for hit in res['hits']['hits']:
    print(hit['_source']['about'])
    print(hit['_score'])
    print('*********************')

#使用聚合功能分析员工的兴趣爱好
res= es.search(index='megacorp',body={
```

```
        "aggs": {
            "all_interests": {
                "terms": { "field": "interests" }
            }
        }
    })
```

从指定位置返回结果:

```
#设置 size 选项,确定要返回的搜索命中数
test=es.search(index=['test'], size=1000, from_=0)
```

使用函数 enumerate()返回结果:

```
result = elastic_client.search(index="some_index", body=query_body, size=999)
all_hits = result['hits']['hits']

#see how many "hits" it returned using the len() function
print ("total hits using 'size' param:", len(result["hits"]["hits"]))

#iterate the nested dictionaries inside the ["hits"]["hits"] list
for num, doc in enumerate(all_hits):
    print ("DOC ID:", doc["_id"], "--->", doc, type(doc), "\n")

    #Use 'iteritems()` instead of 'items()' if using Python 2
    for key, value in doc.items():
        print (key, "-->", value)

    #print a few spaces between each doc for readability
    print ("\n\n")
```

接下来介绍 Elasticsearch DSL 的使用。安装模块:

```
pip install elasticsearch-dsl
```

直接写成一个 dict 的典型的搜索请求如下:

```
from elasticsearch import Elasticsearch
client = Elasticsearch()

response = client.search(
    index="my-index",
    body={
      "query": {
        "bool": {
          "must": [{"match": {"title": "python"}}],
          "must_not": [{"match": {"description": "beta"}}],
          "filter": [{"term": {"category": "search"}}]
        }
      },
```

```
        "aggs" : {
          "per_tag": {
            "terms": {"field": "tags"},
            "aggs": {
              "max_lines": {"max": {"field": "lines"}}
            }
          }
        }
      }
    }
)

for hit in response['hits']['hits']:
    print(hit['_score'], hit['_source']['title'])

for tag in response['aggregations']['per_tag']['buckets']:
    print(tag['key'], tag['max_lines']['value'])
```

这种方法的问题是它非常冗长，容易出现语法错误，例如错误的嵌套、难以修改（如添加另一个过滤器）。

用 Python DSL 重写示例：

```
from elasticsearch import Elasticsearch
from elasticsearch_dsl import Search

client = Elasticsearch()

#将术语查询放在布尔查询的过滤器上下文中
s = Search(using=client, index="my-index") \
    .filter("term", category="search") \
    .query("match", title="python")   \
    .exclude("match", description="beta")

s.aggs.bucket('per_tag', 'terms', field='tags') \
    .metric('max_lines', 'max', field='lines')

response = s.execute()

for hit in response:
    print(hit.meta.score, hit.title)

for tag in response.aggregations.per_tag.buckets:
    print(tag.key, tag.max_lines.value)
```

用一个简单的 Python 类，代表博客系统中的一篇文章：

```
from datetime import datetime
from elasticsearch_dsl import Document, Date, Integer, Keyword, Text, connections
```

```python
#定义一个默认的Elasticsearch客户端
connections.create_connection(hosts=['localhost'])

class Article(Document):
    title = Text(analyzer='snowball', fields={'raw': Keyword()})
    body = Text(analyzer='snowball')
    tags = Keyword()
    published_from = Date()
    lines = Integer()

    class Index:
        name = 'blog'
        settings = {
          "number_of_shards": 2,
        }

    def save(self, ** kwargs):
        self.lines = len(self.body.split())
        return super(Article, self).save(** kwargs)

    def is_published(self):
        return datetime.now() > self.published_from

#在elasticsearch中创建映射
Article.init()

#创建并保存文章
article = Article(meta={'id': 42}, title='Hello world!', tags=['test'])
article.body = ''' looong text '''
article.published_from = datetime.now()
article.save()

article = Article.get(id=42)
print(article.is_published())

#显示群集运行状况
print(connections.get_connection().cluster.health())
```

接下来介绍摄取处理器插件的使用。

首先安装 attachment 插件：

```
#sudo bin/elasticsearch-plugin install ingest-attachment
```

然后重新启动 Elasticsearch，让插件生效。

在 Python 客户端创建一个管道，然后使用摄取处理器插件 attachment，如下所示：

```
from elasticsearch import Elasticsearch
es = Elasticsearch()
body = {
```

```
        "description" : "Extract attachment information",
        "processors" : [
         {
            "attachment" : {
               "field" : "data"
             }
         }
        ]
     }
es.index(index='_ingest', doc_type='pipeline', id='attachment', body=body)
```

1.3.3 定义索引结构

要将 JSON 数据传递给方法的 body 参数，用于创建 Elasticsearch 映射的 Python 字典，代码如下：

```
mapping = {
    "mappings": {
        "properties": {
            "some_string": {
                "type": "text" #formerly "string"
            },
            "some_bool": {
                "type": "boolean"
            },
            "some_int": {
                "type": "integer"
            },
            "some_more_text": {
                "type": "text"
            }
        }
    }
}
```

可以更改分片和副本设置：

```
#mapping dictionary that contains the settings and
#_mapping schema for a new Elasticsearch index:
mapping = {
    "settings": {
        "number_of_shards": 2,
        "number_of_replicas": 1
    },
    "mappings": {
        "properties": {
            "some_string": {
```

```
            "type": "text" #formerly "string"
        },
        "some_bool": {
            "type": "boolean"
        },
        "some_int": {
            "type": "integer"
        },
        "some_more_text": {
            "type": "text"
        }
    }
  }
}
```

index.create()方法的 API 调用的基本布局如下:

```
#create an index with the mapping passed to the 'body' parameter
indices.create(
    index="__INDEX_NAME_GOES_HERE__",
    body=MAPPING_DICT_GOES_HERE
)
```

调用该方法时,只有两个必需的参数:一个是以字符串形式传递到索引的索引名称;另一个是将 Python 字典传递给方法的 body 参数。但是,有一个选项可以指示 Elasticsearch 忽略指定的 HTTP 错误代码。

```
#make an API call to the Elasticsearch cluster
#and have it return a response:
response = elastic_client.indices.create(
    index="some_new_index",
    body=mapping,
    ignore=400 #ignore 400 already exists code
)

#print out the response:
print ('response:', response)
```

如果执行得当,Elasticsearch 集群应返回一个 Python 字典。该字典在响应中具有值为"True"的"acknowledged"键,如下所示:

```
response: {'acknowledged': True, 'shards_acknowledged': True, 'index': 'some_new_index'}
```

评估命令的 Python 条件语句如下所示:

```
if 'acknowledged' in response:
    if response['acknowledged'] == True:
        print ("INDEX MAPPING SUCCESS FOR INDEX:", response['index'])
```

```python
#catch API error response
elif 'error' in response:
    print ("ERROR:", response['error']['root_cause'])
    print ("TYPE:", response['error']['type'])
```

使用 Kibana 控制台或 curl 请求可以验证索引及其映射是否已正确创建。

定义索引结构的完整 Python 代码如下：

```python
#!/usr/bin/env python3
#-*- coding: UTF-8 -*-

from elasticsearch import Elasticsearch
elastic = Elasticsearch(hosts=["localhost"])
#or: elastic = Elasticsearch(hosts=["localhost"])

#mapping dictionary that contains the settings and
#_mapping schema for a new Elasticsearch index:
mapping = {
    "settings": {
        "number_of_shards": 2,
        "number_of_replicas": 1
    },
    "mappings": {
        "properties": {
            "some_string": {
                "type": "text" #formerly "string"
            },
            "some_bool": {
                "type": "boolean"
            },
            "some_int": {
                "type": "integer"
            },
            "some_more_text": {
                "type": "text"
            }
        }
    }
}

#make an API call to the Elasticsearch cluster
#and have it return a response:
response = elastic.indices.create(
```

```
        index="some_new_index",
        body=mapping,
        ignore=400 #ignore 400 already exists code
)

if 'acknowledged' in response:
    if response['acknowledged'] == True:
        print ("INDEX MAPPING SUCCESS FOR INDEX:", response['index'])

#catch API error response
elif 'error' in response:
    print ("ERROR:", response['error']['root_cause'])
    print ("TYPE:", response['error']['type'])

#print out the response:
print ('\nresponse:', response)
```

Elasticsearch 不会分析 Keyword 数据类型，这意味着索引的字符串将保持不变。Keyword 数据类型可用于聚合列。

与 Keyword 字段数据类型不同，索引到 text 类型的字符串在存储到反向索引中之前将经过分析器处理。默认情况下，Elasticsearch 的标准分析器将拆分并小写化索引的字符串。

第 2 章

Python 技术基础

本章介绍开发 Web 应用所用到的 Python 技术基础。

2.1 变量

在 Python 中定义变量时,不需要声明类型,但变量在内部是有类型的。例如,在交互式环境输入如下代码,即输出变量 a 的类型:

```
>>> a='test'
>>> type(a)    #调用函数 type()得到变量 a 的类型
<class 'str'>
```

2.2 注释

与 Shell 类似,Python 脚本中用#表示注释。但是,如果#位于第一行开头,并且是#!(称为 Shebang),则例外。它表示该脚本使用后面指定的解释器/usr/bin/python3 解释执行。每个脚本程序只能在开头包含这条语句。

为了能够在源代码中添加中文注释,需要把源代码保存为 UTF-8 格式。例如:

```
#-*- coding: UTF-8 -*

import spacy
en_nlp = spacy.load('en')    #加载英文模型
```

2.3 简单数据类型

本节介绍包括数值、字符串和数组在内的简单数据类型。

2.3.1 数值

Python 中有三种不同的数值类型：int（整数）、float（浮点数）和 complex（复数）。与 Java 或者 C 语言中的 int 类型不同，Python 中的 int 类型是无限精度的。例如：

```
>>> i=3243244444444444444444444444444444444444448797687567567657000000000000000000000000000000000000000000000000000000000000000000000000564564
>>> i
3243244444444444444444444444444444444444448797687567567657000000000000000000000000000000000000000000000000000000000000000000000000564564
>>> type(i)
<class 'int'>
```

Python 依据 IEEE 754 标准使用二进制表示 float（浮点数），所以存在表示精度的问题。例如：

```
>>> 0.1 == 0.10000000000000000000001
True
```

可以使用 decimal 模块采用十进制表示完整的小数。例如：

```
>>> import decimal
>>> a = decimal.Decimal('0.1')
>>> b = decimal.Decimal('0.10000000000000000000001')
>>> a == b
False
```

在傅里叶变换中会用到复数。复数在 Python 中是一个基本数据类型（complex）。例如：

```
>>> complex(2,3)
(2+3j)
```

一个复数有一些内置的访问器：

```
>>> z = 2+3j
>>> z.real
2.0
>>> z.imag
3.0
>>> z.conjugate()
(2-3j)
```

几个内置函数支持复数：

```
>>> abs(3 + 4j)
5.0
>>> pow(3 + 4j, 2)
(-7+24j)
```

标准模块 cmath 具有处理复数的更多功能：

```
>>> import cmath
```

```
>>> cmath.sin(2 + 3j)
(9.15449914691143-4.168906959966565j)
```

用于数值运算的算术运算符说明见表 2-1。

表 2-1 算术运算符

语 法	数 学 含 义	运算符名字
a+b	$a+b$	加
a-b	$a-b$	减
a*b	$a \times b$	乘法
a/b	$a \div b$	除法
a//b	$\lfloor a \div b \rfloor$	地板除
a%b	$a \bmod b$	模
-a	$-a$	取负数
abs(a)	$\lvert a \rvert$	绝对值
a**b	a^b	指数
math.sqrt(a)	$\sqrt{a}$	平方根

对于"/"运算，就算分子和分母都是 int，返回的也将是浮点数。例如：

```
>>> print(1/3)
0.3333333333333333
```

Python 支持不同的数字类型相加，它使用数字类型强制转换的方式来解决数字类型不一致的问题。也就是说，它会将一个操作数转换成与另一个操作数相同的数据类型。

如果有一个操作数是复数，则另一个操作数被转换为复数：

```
>>> 3.0 + (5+6j)          #非复数转换为复数
(8+6j)
```

整数转换为浮点数：

```
>>> 6 + 7.0               #非浮点型转换为浮点型
13.0
```

Python 代码中一般一行就是一条语句，但是可以使用斜杠（\）将一条语句分为多行显示。例如：

```
>>> a = 1
>>> b = 2
>>> c = 3
>>> total = a + \
... b + \
... c
```

```
>>> total
6
```

2.3.2 字符串

使用 strip() 方法可以去掉字符串首尾的空格或者指定的字符。例如：

```
term = "   hi       ";
print(term.strip());           #去除首尾空格
```

使用 split() 方法将句子分成单词。例如下面的代码中，mary 是一个单一的字符串，尽管这是一个句子，但这些词语并没有表示成严谨的单位。为此，需要一种不同的数据类型：字符串列表，其中每个字符串对应一个单词。使用 split() 方法把句子切分成单词：

```
>>> mary = 'Mary had a little lamb'
>>> mary.split()
['Mary', 'had', 'a', 'little', 'lamb']
```

split() 方法根据空格拆分 mary，返回的结果是 mary 中的单词列表。此列表包含函数 len() 演示的 5 个项目。对于 mary，函数 len() 返回字符串中的字符数（包括空格）。

```
>>> mwords = mary.split()
>>> mwords
['Mary', 'had', 'a', 'little', 'lamb']
>>> len(mwords)                #mwords 中的项目数
5
>>> len(mary)                  #字符数
22
```

空白字符包括空格' '，换行符'\n'和制表符'\t'等。.split() 分隔这些字符的任何组合序列：

```
>>> chom = ' colorless    green \n\tideas\n'
>>> print(chom)
 colorless    green
    ideas

>>> chom.split()
['colorless', 'green', 'ideas']
```

通过提供可选参数，.split('x') 可用于在特定子字符串'x'上拆分字符串。如果没有指定'x'，则.split() 只是在所有空格上分割，例如：

```
>>> mary = 'Mary had a little lamb'
>>> mary.split('a')                    #根据'a'切分
['M', 'ry h', 'd ', ' little l', 'mb']
>>> hi = 'Hello mother,\nHello father.'
>>> print(hi)
Hello mother,
```

```
Hello father.
>>> hi.split()                    #没有给出参数：在空格上分割
['Hello', 'mother,', 'Hello', 'father.']
>>> hi.split('\n')                #仅在'\n'上分割
['Hello mother,', 'Hello father.']
```

但是，如果想将一个字符串拆分成一个字符列表呢？在 Python 中，字符只是长度为 1 的字符串。函数 list() 将字符串转换为单个字母的列表：

```
>>> list('hello world')
['h', 'e', 'l', 'l', 'o', ' ', 'w', 'o', 'r', 'l', 'd']
```

如果有一个单词列表，可以使用.join()方法将它们重新组合成一个单独的字符串。在"分隔符"字符串'x'上调用，'x'.join(y)连接列表 y 中由'x'分隔的每个元素。例如下面的代码，mwords 中的单词用空格连接回句子字符串：

```
>>> mwords
['Mary', 'had', 'a', 'little', 'lamb']
>>> ' '.join(mwords)
'Mary had a little lamb'
```

也可以在空字符串""上调用该方法作为分隔符。效果是列表中的元素连接在一起，元素之间没有任何内容。例如下面的代码，将一个字符列表放回到原始字符串中：

```
>>> hi = 'hello world'
>>> hichars = list(hi)
>>> hichars
['h', 'e', 'l', 'l', 'o', ' ', 'w', 'o', 'r', 'l', 'd']
>>> ''.join(hichars)
'hello world'
```

对一个字符串取子串的例子代码如下：

```
>>> x = "Hello World!"
>>> x[2:]
'llo World!'
>>> x[:2]
'He'
>>> x[:-2]
'Hello Worl'
>>> x[-2:]
'd!'
>>> x[2:-2]
'llo Worl'
```

使用函数 ord() 和 chr() 实现字符串和整数之间的相互转换：

```
>>> a = 'v'
>>> i = ord(a)
>>> chr(i)
```

```
'v'
```

2.3.3 数组

创建一个数组，然后向这个数组添加元素的代码如下：

```
>>> temp_list = []
>>> print(temp_list)
[]
>>> temp_list.append("one")
>>> temp_list.append("two")
>>> print(temp_list)
['one', 'two']
>>>
```

创建一个指定长度的数组：

```
>>> size = 10
>>> lst = [None] * size
>>> lst
[None, None, None, None, None, None, None, None, None, None]
```

2.4 字面值

Python 包括以下几种类型的字面值。
- 数字：整数、浮点数、复数。
- 字符串：以单引号、双引号或者三引号定义字符串。
- 布尔值：True 和 False。
- 空值：None。

Python 中有 4 种不同的字面值集合：列表字面值、元组字面值、字典字面值和集合字面值。示例代码如下：

```
fruits = ["apple", "mango", "orange"]            #列表
numbers = (1, 2, 3)                              #元组
alphabets = {'a':'apple', 'b':'ball', 'c':'cat'} #字典
vowels = {'a', 'e', 'i', 'o', 'u'}               #集合

print(fruits)
print(numbers)
print(alphabets)
print(vowels)
```

2.5 控制流

完成一件事情要有流程控制。例如，洗衣服有三个步骤：把脏衣服放进洗衣机，等洗衣机洗好衣服，晾衣服。这是顺序控制结构。

顺序执行的代码采用相同的缩进，叫作一个代码块。Python 没有像 Java 或者 C#语言那样采用{}分隔代码块，而是采用代码缩进和冒号来区分代码之间的层次。

缩进的空格数量是可变的，但是所有代码块语句必须包含相同的缩进空格数量。NodePad++这样的文本编辑器支持选择多行代码后，按 Tab 键可以改变代码块的缩进格式。

控制流用来根据运行时的情况调整语句的执行顺序。流程控制语句可以分为条件语句和迭代语句。

2.5.1 if 语句

路径不存在，就创建它，可以使用条件语句实现。条件语句的一般形式如下：

```
if 条件:
    语句1
    语句2
    ...
elif 条件:
    语句1
    语句2
    ...
else:
    语句1
    语句2
    ...
语句x
```

例如，判断一个数是否是正数的代码如下：

```
x = -32.2;
isPositive = (x > 0);
if isPositive:
    print(x, " 是正数");
else:
    print(x, " 不是正数");
```

这里的 if 复合语句，首行以关键字开始，以冒号（:）结束。

使用关系运算符和条件运算符作为判断依据。关系运算符返回一个布尔值。关系运算符的说明见表 2-2。

表 2-2 关系运算符

运算符	用法	返回 true, 如果……
>	a > b	a 大于 b
>=	a >= b	a 大于或等于 b

续表

运 算 符	用 法	返回 true，如果……
<	a < b	a 小于 b
<=	a <= b	a 小于或等于 b
==	a == b	a 等于 b
!=	a != b	a 不等于 b

如果针对多个值测试一个变量，则可以在 if 条件判断中使用一个集合：

```
x = "Wild things"
y = "throttle it back"
z = "in the beginning"
if "Wild" in {x, y, z}: print (True)
```

2.5.2 循环

使用复印机复印一个证件，可以设定复制的份数。例如，复制 3 份。在 Python 中，可以使用 for 循环或者 while 循环实现多次重复执行一个代码块。

for 循环可以遍历任何序列。例如，输出数组中的元素：

```
mylist = [1,2,3]
for item in mylist:
    print(item)
```

输出字符串中的字符：

```
>>> for c in '风调雨顺' :
...     print(c,type(c))
...
风 <class 'str'>
调 <class 'str'>
雨 <class 'str'>
顺 <class 'str'>
```

或者借助函数 range()遍历字符串中的字符：

```
>>> word = '风调雨顺'
>>> for i in range(len(word)):
...     print(word[i])
...
风
调
雨
顺
```

因为 Python 3 中并不存在表示单个字符的数据类型，所以返回的变量 c 仍然是 str 类型。

输出字符串'banana'中每个出现的字符及其位置：

```
>>> for c in enumerate('banana'):
...     print(c)
...
(0, 'b')
(1, 'a')
(2, 'n')
(3, 'a')
(4, 'n')
(5, 'a')
```

在 Python 中，可以将一个可选的"else"块与循环关联。"else"块仅在循环完成所有迭代后才执行。例如：

```
for val in range(5):
    print(val)
else:
    print("The loop has completed execution")
```

输出：

```
0
1
2
3
4
The loop has completed execution
```

每一次在执行循环代码块之前，根据循环条件决定是否继续执行循环代码块，当满足循环条件时，继续执行循环体中的代码。在循环条件之前写上关键词 while。这里的 while 就是"当"的意思。例如，当用户直接按 Enter 键时退出循环：

```
import sys

while True:
    line = sys.stdin.readline().strip()
    if not line:
        break
    print(line)
```

2.6 列表

使用一个列表可以存储任何类型的对象：

```
list1 = ['physics', 'chemistry', 1997, 2000];
print("list1[0]: ", list1[0])
```

输出：

```
list1[0]: physics
```

可以通过切片运算来获得一个列表：

```
>>> a = [0, 1, 2, 3, 4, 5, 6, 7, 8, 9]
>>> a[1:9]
[1, 2, 3, 4, 5, 6, 7, 8]
```

应用 lambda 表达式截短字符串：

```
>>> lines = ['this', 'is', 'a', 'list', 'of', 'words']
>>> list(map(lambda it: it[:3], lines))
['thi', 'is', 'a', 'lis', 'of', 'wor']
```

2.7 元组

元组是一个不可变的 Python 对象序列。元组变量的赋值要在定义时就进行，定义时赋值后就不允许有修改。

```
tup1 = ('physics', 'chemistry', 1997, 2000);
tup2 = (1, 2, 3, 4, 5, 6, 7 );
print( "tup1[0]: ", tup1[0]);
print( "tup2[1:5]: ", tup2[1:5]);
```

通常将元组用于异构（不同）数据类型，将列表用于同类（相似）数据类型。

包含多个项目的文字元组可以分配给单个对象。当发生这种情况时，就好像元组中的项目已经"打包"到对象中。

```
>>> t = ('foo', 'bar', 'baz', 'qux')
```

将元组中的元素分别赋给变量称为拆包。

```
>>> (s1, s2, s3, s4) = t
>>> s1
'foo'
>>> s2
'bar'
>>> s3
'baz'
>>> s4
'qux'
```

包装和拆包可以合并为一条语句，以进行复合分配：

```
>>> (s1, s2, s3, s4) = ('foo', 'bar', 'baz', 'qux')
>>> s1
'foo'
```

```
>>> s2
'bar'
>>> s3
'baz'
>>> s4
'qux'
```

可以构建一个由元组组成的数组：

```
>>> pairs = [("a", 1), ("b", 2), ("c", 3)]
>>> for a, b in pairs:
...     print(a, b)
...
a 1
b 2
c 3
```

使用命名元组可以给元组中的元素起一个有意义的名字：

```
import collections

#声明一个名为 Person 的命名元组，这个元组包含 name 和 age 两个键
Person = collections.namedtuple('Person', 'name age')

#使用命名元组
bob = Person(name='Bob', age=30)
print('\nRepresentation:', bob)

jane = Person(name='Jane', age=29)
print('\nField by name:', jane.name)

print('\nFields by index:')
for p in [bob, jane]:
    print('{} is {} years old'.format(*p))
```

2.8 集合

使用 in 运算符可以检查给定元素是否存在于集合中。如果集合中存在指定元素，则返回 True；否则，返回 False。

```
>>> s = {1,2,3,4,5}          #创建 set 对象并将其分配给变量 s
>>> contains = 1 in s        #判断是否包含的例子
>>> print(contains)
True
>>> contains = 6 in s
>>> print(contains)
```

```
False
```

输出字符串'banana'中的字符集合：

```
>>> set(c for (i,c) in enumerate('banana'))
{'n', 'a', 'b'}
```

可以使用 set.update()方法增加项目到集合。

```
>>> A = [1, 2, 3]
>>> S = set()
>>> S.update(A)
>>> S
{1, 2, 3}
```

可以使用 set.intersection()方法找出两个集合都包含的元素。

```
A = {2, 3, 5, 4}
B = {2, 5, 100}

print(B.intersection(A))    #输出2,5
```

2.9　字典

使用字典可以存取键/值对。若访问字典元素，则可以使用熟悉的方括号和键来获取字典中的值。

```
dict = {'Name': 'Zara', 'Age': 7, 'Class': 'First'}
print("dict['Name']: ", dict['Name'])
print("dict['Age']: ", dict['Age'])
```

通过为该键指定值可以在字典上创建新的键/值对。如果该键不存在，则将该键/值对添加到字典。如果该键已经存在，则覆盖它指向的当前值。

```
d = {'key':'value'}
print(d)          #输出 {'key': 'value'}
d['mynewkey'] = 'mynewvalue'
print(d)          #输出 {'mynewkey': 'mynewvalue', 'key': 'value'}
```

运行如下语句会抛出 KeyError 异常：

```
print(d['noexists'])
```

运行如下语句会返回 None：

```
print(d.get('noexists'))
```

如果只需要判断键是否在字典中存在，则可以使用 in 关键字实现：

```
>>> d = {'a': 1, 'b': 2}
>>> 'a' in d  #<==评估为 True
```

```
True
>>> 'c' in d  #<==评估为 False
False
```

键可以是复杂数据类型：

```
>>> transProb = {}
>>> transProb[('yes','是')] = 0.6
>>> transProb[('yes','好')] = 0.4
>>> transProb[('good','好')]                     #访问不存在的键触发异常
Traceback (most recent call last):
  File "<stdin>", line 1, in <module>
KeyError: ('good', '好')
```

为了避免异常，可以通过函数 collections.defaultdict() 设置默认值：

```
>>> from decimal import Decimal
>>> import collections
>>> transProb = collections.defaultdict(Decimal)  #设定默认值
>>>
>>> transProb[('yes','是')] = 0.6
>>> transProb[('yes','好')] = 0.4
>>>
>>> transProb[('good','好')]                      #返回默认值
Decimal('0')
```

如果需要根据字典中的值排序，由于字典本质上是无序的，则可以把排序结果保存到有序的列表。

```
>>> x = {1: 2, 3: 4, 4: 3, 2: 1, 0: 0}
>>> sorted_by_value = sorted(x.items(), key=lambda kv: kv[1])
>>> print(sorted_by_value)
[(0, 0), (2, 1), (1, 2), (4, 3), (3, 4)]
```

OrderedDict 是一个字典子类，它会记住键/值对的顺序。

```
import collections

print('普通的字典:')
d = {}
d['a'] = 'A'
d['b'] = 'B'
d['c'] = 'C'

for k, v in d.items():
    print(k, v)

print('\n有序的字典:')
d = collections.OrderedDict()
d['a'] = 'A'
```

```
d['b'] = 'B'
d['c'] = 'C'
d['a'] = 'a'

for k, v in d.items():
    print(k, v)
```

2.10 位数组

位数组（也称为位图）通常用于快速内存访问上的一种数据结构。位图可以表示文档的切分结果。不幸的是，位图可能会使用太多内存。为了补偿，可以使用压缩位图。

PyRoaringBitMap（https://github.com/Ezibenroc/PyRoaringBitMap）是一个 C 语言库 CRoaring 的 Python 包装器。

可以使用 PyPi 安装 pyroaring：

```
#pip3 install pyroaring
```

或者从 whl 文件安装：

```
#pip3 install --user https://github.com/Ezibenroc/PyRoaringBitMap/releases/download/0.2.1/pyroaring-0.2.1-cp36-cp36m-linux_x86_64.whl
```

几乎可以像使用经典的 Python 集合那样在代码中使用 BitMap：

```
from pyroaring import BitMap
bm1 = BitMap()
bm1.add(3)
bm1.add(18)
bm2 = BitMap([3, 27, 42])
print("bm1       = %s" % bm1)
print("bm2       = %s" % bm2)
print("bm1 & bm2 = %s" % (bm1&bm2))    #按位运算
print("bm1 | bm2 = %s" % (bm1|bm2))
```

输出：

```
bm1       = BitMap([3, 18])
bm2       = BitMap([3, 27, 42])
bm1 & bm2 = BitMap([3])
bm1 | bm2 = BitMap([3, 18, 27, 42])
```

遍历位数组：

```
>>> a = iter(bm1)                      #取得iterator
>>> print(next(a, None))               #取得下一个元素,如果没有,则返回None
3
>>> print(next(a, None))
```

```
18
>>> print(next(a, None))
None
```

2.11 模块

使用 import 语句可以导入一个在.py 文件中定义的函数。一个.py 文件称为一个模块（Module）。例如，存在一个 re.py 文件，可以使用 import re 语句导入这个正则表达式模块。

使用正则表达式模块去掉一些标点符号的例子代码如下：

```
import re

line = 'Hi.'
normtext = re.sub(r'[\.,:;\?]', '', line)
print(normtext)
```

从 re 模块直接导入 sub 函数的例子代码如下：

```
from re import sub

line = 'Hi.'
normtext = sub(r'[\.,:;\?]', '', line)
print(normtext)
```

模块越来越多以后，会难以管理。例如，可能会出现重名的模块。又如，一个班里有两个叫陈晨的同学，如果他们在不同的小组，可以叫第一组的陈晨或者第三组的陈晨，这样就能区分同名。为了避免名字冲突，模块可以位于不同的命名空间，叫作包。在模块名前面可以加上包名限定，这样即使模块名相同，也不会冲突。

为了查看本地有哪些模块可用，可以在 Python 交互式环境中输入：

```
help('modules')
```

对于大项目，可以把多个模块文件置于同一个目录下组成包，而且必须在该目录下放置一个__init__.py 文件来让 Python 识别出这是一个包。

2.12 函数

把一段多次重复出现的函数命名成一个有意义的名字，然后通过名字来执行这段代码。有名字的代码段就是一个函数。

Python 解释器内置了一些函数。其中，函数 str()将对象转换为易于阅读的字符串；函数 len()用于获取对象的长度；函数 id()返回对象的标识；函数 print()将给定对象打印到标准输出设备（屏幕）或文本流文件。

2.12.1　print 函数

显示某个目录下的文件数量的代码如下：

```
import os

folderlist = os.listdir('/home/soft/kaldi/')
total_num_file = len(folderlist)

print ('total '+total_num_file+' files')
```

这样会出错，因为 Python 不支持+运算中的整数自动转换成字符串。调用函数 str()可以将整数转换成字符串：

```
print ('total '+str(total_num_file)+' files')
```

或者格式化：

```
print ('total have %d files' % (total_num_file))    #%d 表示输出整数
```

另外一种格式化输出的方法是使用 str.format()方法，下面的代码比较了这两种方法：

```
>>> sub1 = "python string!"
>>> sub2 = "an arg"
>>> a = "i am a %s" % sub1
>>> b = "i am a {0}".format(sub1)
>>> print(a)
i am a python string!
>>> print(b)
i am a python string!
>>> c = "with %(kwarg)s!" % {'kwarg':sub2}
>>> print(c)
with an arg!
>>> d = "with {kwarg}!".format(kwarg=sub2)
>>> print(d)
with an arg!
```

如下的代码会出错：

```
>>> name=(1, 2, 3)
>>> print("hi there %s" % name)
Traceback (most recent call last):
  File "<stdin>", line 1, in <module>
TypeError: not all arguments converted during string formatting
```

函数 print()用到的格式化字符串的约定见表 2-3。

表 2-3 格式化字符串中的转换类型

转换类型	含义
d,i	带符号的十进制整数
o	不带符号的八进制数
u	不带符号的十进制数
x	不带符号的十六进制数（小写）
X	不带符号的十六进制数（大写）
e	科学记数法表示的浮点数（小写）
E	科学记数法表示的浮点数（大写）
f,F	十进制浮点数
g	如果指数大于-4 或者小于精度值，则和 e 相同，其他情况和 f 相同
G	如果指数大于-4 或者小于精度值，则和 E 相同，其他情况和 F 相同
C	单字符（接受整数或者单字符字符串）
r	字符串（使用 repr()转换任意 Python 对象）
s	字符串（使用 str()转换任意 Python 对象）

2.12.2 定义函数

使用关键字 def 定义一个函数。例如：

```
def square(number):          #定义一个名为 square 的函数
    return number * number   #返回一个数的平方
print(square(3))             #输出：9
```

代码中可以给函数增加说明：

```
def square_root(n):
    """计算一个数字的平方根。

    Args:
        n: 用来求平方根的数字。
    Returns:
        n 的平方根。
    Raises:
        TypeError: 如果 n 不是数字。
        ValueError: 如果 n 是负数。

    """
    pass
```

参数可以有默认值。例如，定义一个名为 RunKaldiCommand 的函数：

```
import subprocess
```

```
def RunKaldiCommand(command, wait = True):          #wait 的默认值是 True
    """通常执行由管道连接的一系列命令,所以使用shell=True """
    p = subprocess.Popen(command, shell = True,
                    stdout = subprocess.PIPE,
                    stderr = subprocess.PIPE)

    if wait:
        [stdout, stderr] = p.communicate()
        if p.returncode is not 0:                   #执行命令出现错误
            raise Exception("There was an error while running the command {0}\n".format(command)+"-"*10+"\n"+stderr)
        return stdout, stderr
    else:
        return p
```

使用这个函数:

```
RunKaldiCommand("ls -lh")
```

这里只给 RunKaldiCommand()方法的第一个参数传递了值,第二个值采用默认的 True。如果需要声明可变数量的参数,则在这个参数前面加*。例如:

```
def myFun(*argv):
    for arg in argv:
        print (arg)

myFun('Hello', 'a', 'to', 'b')
```

函数定义中的特殊语法**kwargs 用于传递一个键/值对的,且可变长度的参数列表。例如:

```
def myFun(**kwargs):
    for key, value in kwargs.items():
        print ("%s == %s" %(key, value))

#调用函数
myFun(first ='test', mid ='for', last='abc')
```

输出结果如下:

```
first == test
mid == for
last == abc
```

每个 Python 文件/脚本(模块)都有一些未明确声明的内部属性。其中,一个属性是 __builtins__属性,它本身包含许多有用的属性和功能。在这里可以找到__name__属性,根据模块的使用方式,它可以具有不同的值。

当把 Python 模块作为程序直接运行时(无论是从命令行还是双击它),__name__中包含的值都是文字字符串"__main__"。

相比之下，当一个模块被导入到另一个模块中（或者在 Python REPL 被导入）时，__name__属性中的值是模块本身的名称（即隐式声明它的 Python 文件/脚本的名称）。

Python 脚本执行的方式是自上而下的。指令在解释器读取它们时执行。这可能是一个问题，如果想要做的就是导入模块并利用它的一个或两个方法，则可以有条件地执行这些指令——将它们包装在一个 if 语句块中。

这是'main 函数'的目的。它是一个条件块，因此除非满足给定的条件，否则不会处理函数 main()。

函数 main()的例子代码如下：

```
import sys

def main():
    if len(sys.argv) != 2:
        sys.stderr.write("Usage: {0} <min-count>\n".format(sys.argv[0]))
        raise SystemExit(1)

    words = {}
    for line in sys.stdin.readlines():
        parts = line.strip().split()
        words[parts[1]] = words.get(parts[1], 0) + int(parts[0])

    for word, count in words.iteritems():
        if count >= int(sys.argv[1]):
            print ("{0} {1}".format(count, word))

if __name__ == '__main__':
    main()
```

2.13　面向对象编程

定义一个 Token 类描述词在文本中的位置：

```
class Token(object):                              #Token 类是 object 的子类
    """标记"""
    def __init__(self, text, offset, data=None):  #构造方法
        self.offset = offset                      #词在文档中的开始位置
        self.text = text                          #词
        self.end = offset + len(text)             #词在文档中的结束位置
        self.data = data if data else {}

    def set(self, prop, info):
```

```python
        self.data[prop] = info

    def get(self, prop, default=None):
        return self.data.get(prop, default)
```

调用这个构造方法来创造对象。例如，有个词出现在文档的开始位置：

```python
t = Token("剧情", 0)   #出现在开始位置的"剧情"这个词
```

例如，有个根据指定字符分割输入字符串的 StringTokenizer 类实现如下：

```python
class StringTokenizer(object):
    """字符串分隔类"""

    def __init__(self,text:str, delim:str):
        self.currentPosition = 0
        self.newPosition = -1
        self.text = text
        self.maxPosition = len(text)
        self.delimiters = delim

    def skip_delimiters(self,startPos:int)->int:
        """跳过分隔符"""
        position = startPos
        while ( position < self.maxPosition):
            c = self.text[position]
            if( self.delimiters.find(c) == -1 ):
                break
            position+=1
        return position

    def scan_token(self,startPos:int)->int:
        """扫描符号"""
        position = startPos
        while (position < self.maxPosition):
            c = self.text[position]
            if( self.delimiters.find(c) != -1 ):
                break
            position+=1
        return position

    def next_token(self):
        """取得下一个标记"""
        self.currentPosition = self.skip_delimiters(self.currentPosition)
        start = self.currentPosition
        self.currentPosition = self.scan_token(self.currentPosition)
        return self.text[start: self.currentPosition]
```

使用这个 StringTokenizer 类：

```
st = StringTokenizer("2#7#8#9#道","#")
print(st.next_token())
print(st.next_token())
print(st.next_token())
print(st.next_token())
print(st.next_token())
```

2.13.1 静态方法

静态方法是一种属于类中的函数，同时表明它不需要访问类。@classmethod 创建了一个方法，其第一个参数是从中调用的类（而不是类实例），@staticmethod 没有任何隐式参数。例如：

```
class A(object):
    def foo(self, x):
        print( "executing foo(%s, %s)" % (self, x) )

    @classmethod
    def class_foo(cls, x):
        print( "executing class_foo(%s, %s)" % (cls, x) )

    @staticmethod
    def static_foo(x):
        print( "executing static_foo(%s)" % x )
a = A()
```

下面是对象实例调用方法的常用方法。对象实例 a 作为第一个参数隐式传递。

```
a.foo(1)
#executing foo(<__main__.A object at 0xb7dbef0c>,1)
```

使用类可以调用 class_foo。

```
A.class_foo(1)
#executing class_foo(<class '__main__.A'>,1)
```

对于静态方法，self（对象实例）和 cls（类）都不会作为第一个参数隐式传递。调用静态方法：

```
A.static_foo('hi')
#executing static_foo(hi)
```

在类定义中声明而不是在方法内部声明的变量是静态变量：

```
class MyClass:
    i = 3
```

```
print(MyClass.i)            #输出静态变量 i 的值

m = MyClass()
m.i = 4                     #实例变量
print(MyClass.i, m.i)       #输出静态变量和实例变量的值 3 4
```

生成唯一 ID:

```
import itertools

class BarFoo:

    id_iter = itertools.count()

    def __init__(self):
        self.id = next(self.id_iter)
```

2.13.2 __call__方法

Python 有一组内置方法，而__call__是其中之一。__call__方法可以使 Python 程序员编写实例的行为类似于函数的类。当实例作为函数调用时，如果定义了此方法，则 x(arg1, arg2, ...)是 x.__call__(arg1, arg2, ...)的简写。

object()是 object.__call__()的简写。

示例 1：

```
class Example:
    def __init__(self):
        print("Instance Created")

    #Defining __call__ method
    def __call__(self):
        print("Instance is called via special method")

#Instance created
e = Example()

#__call__ method will be called
e()
```

输出：

```
Instance Created
Instance is called via special method
```

示例 2：

```
class Product:
    def __init__(self):
```

```
        print("Instance Created")

    #Defining __call__ method
    def __call__(self, a, b):
        print(a * b)

#Instance created
ans = Product()

#__call__ method will be called
ans(10, 20)
```

输出:

```
Instance Created
200
```

2.14 使用 StringIO 模块

StringIO 模块是内存中类似文件的对象。创建 StringIO 对象时，可通过将字符串传递给构造函数来对其进行初始化。如果未传递任何字符串，则 StringIO 将以空开始。在这两种情况下，文件上的初始光标都从零开始。

例子代码如下：

```
from io import StringIO

#The arbitrary string.
string = 'Hello and welcome to GeeksForGeeks.'

#Using the StringIO method
#to set as file object.
file = StringIO(string)

#Reading the initial file:
print(file.read())

#To set the cursor at 0.
file.seek(0)

#This will drop the file after
#index 18.
file.truncate(18)

#File after truncate.
```

```
print(file.read())
file.seek(0)
print(file.read())
```

使用 StringIO 实现的字符缓存：

```
import io

class StringBuilder(object):

    def __init__(self):
        self._stringio = io.StringIO()

    def __str__(self):
        return self._stringio.getvalue()

    def append(self, *objects, sep=' ', end=''):
        print(*objects, sep=sep, end=end, file=self._stringio)

sb = StringBuilder()
sb.append('a')
sb.append('b', end='\n')
sb.append('c', 'd', sep=',', end='\n')
print(sb)   #'ab\nc,d\n'
```

2.15 文件操作

文件的绝对路径由目录和文件名两部分构成，示例代码如下：

```
import os.path

path = '/home/data/file.wav'

print(os.path.abspath(path))      #返回绝对路径（包含文件名的全路径）
print(os.path.basename(path))     #返回路径中包含的文件名
print(os.path.dirname(path))      #返回路径中包含的目录
```

输出：

```
/home/data/file.wav
file.wav
/home/data
```

2.15.1 读写文件

调用函数 open(fileName)返回一个_io.TextIOWrapper 对象。例如，文本文件 a.txt 包含

以下内容:

```
the quick person did not realize his speed and the quick person bumped
```

统计文本中的词频:

```
import re
from collections import Counter
words = re.findall('\w+', open('a.txt').read())
print( Counter(words) )
```

输出如下:

```
Counter({'the': 2, 'quick': 2, 'person': 2, 'did': 1, 'not': 1, 'realize': 1, 'his': 1, 'speed': 1, 'and': 1, 'bumped': 1})
```

逐行读入文本文件:

```
lexicon = open("lexicon.txt")

for line in lexicon:
    line = line.strip()
    print(line,"\n")

lexicon.close()
```

读入 UTF-8 编码格式的文本文件:

```
import codecs
import sys

transcript = codecs.open(sys.argv[1], "r", "UTF-8")     #第一个参数传入文件名

for line in transcript:
    print(line)

transcript.close()
```

为了实现写入文本文件,可以使用'w'模式的函数 open()以写模式打开新文件。

```
new_path = "a.speaker_info"
fout = open(new_path,'w')
```

需要注意,如果 new_days.txt 在打开文件前已经存在,它的旧内容将被破坏,所以在使用'w'模式时要小心。

一旦打开新文件,可以使用写入操作<file>.write()将数据放入文件中。写入操作接受单个参数,该参数必须是字符串,并将该字符串写入文件。如果要在文件中开始新行,则必须明确提供换行符。例子代码如下:

```
fout.write("\nID:\t1212")
```

关闭文件可确保磁盘上的文件和文件变量之间的连接已完成。关闭文件还可确保其他

程序能够访问它们并保证数据安全。所以，一定要确保关闭文件。现在，使用<file>.close()函数关闭所有文件。

```
fout.close()
```

使用函数 open()创建文件对象可以采用的模式总结如下：
- 'r'用于读取现有文件（默认值；可以省略）。
- 'w'用于创建写入的新文件。
- 'a'用于将新内容附加到现有文件。

取得文件的修改时间：

```
>>> import os
>>> os.stat('test.txt').st_mtime
1559512015.9773836
```

对于 JSON 格式的文件，可以导入 json 模块读取：

```
import json
data = json.load(open('my_file.json', 'r'))
```

演示 JSON 文件的内容如下：

```
{"hello":"lietu"}
```

演示读取 JSON 格式的文件如下：

```
>>> import json
>>> print(json.load(open('my_file.json','r')))
{u'hello': u'lietu'}
```

2.15.2　重命名文件

使用 os.rename()方法可以重命名文件。首先用 touch 命令创建一个空文件：

```
#touch ./test1
```

然后把 test1 重命名为 test2：

```
import os
src= 'test1'
dst= 'test2'
os.rename(src, dst)
```

2.15.3　遍历文件

使用 os.scandir()方法遍历一个目录。os.scandir()方法返回一个迭代器。

```
import os

with os.scandir('/home/') as entries:
```

```
        for entry in entries:
            print(entry.name)
```

这里通过 with 语句使用上下文管理器关闭迭代器并在迭代器耗尽后自动释放获取的资源。

只打印出一个目录下的文件：

```
dir_entries = os.scandir('/home/')
for entry in dir_entries:
    if entry.is_file():                    #判断项目是否为文件
        print(f'{entry.name}')
```

如果要遍历一个目录树并处理树中的文件，则可以使用 os.walk()方法。os.walk()方法默认以自上而下的方式遍历目录：

```
import os
for root, dirs, files in os.walk("/home/"):
    for name in files:
        print(os.path.join(root, name))    #打印文件
    for name in dirs:
        print(os.path.join(root, name))    #打印目录
```

2.16 迭代器

迭代器（iterator）用来遍历生成器中的元素。通过函数 next()迭代 iterator 对象中的元素。

```
>>> r = range(5)
>>> itr = iter(r)
>>> next(itr)
0
```

如果没有元素，则抛出 StopIteration 异常。为了避免函数 next()抛出 StopIteration 异常，可以指定一个默认值。

```
>>> print(next(itr, None))     #取得下一个元素,如果没有,则返回None
1
```

支持迭代的生成器 StringTokenizer 类需要实现__iter__()方法。

```
class StringTokenizer(object):
    """字符串分隔类"""

    def __init__(self,text:str, delim:str):
        self.currentPosition = 0
        self.newPosition = -1
        self.text = text
        self.maxPosition = len(text)
```

```python
        self.delimiters = delim

    def skip_delimiters(self,startPos:int)->int:
        """跳过分隔符"""
        position = startPos
        while ( position < self.maxPosition):
            c = self.text[position]
            if( self.delimiters.find(c) == -1 ):
                break
            position+=1
        return position

    def scan_token(self,startPos:int)->int:
        """扫描标记"""
        position = startPos
        while (position < self.maxPosition):
            c = self.text[position]
            if( self.delimiters.find(c) != -1 ):
                break
            position+=1
        return position

    def next_token(self):
        """取得下一个标记"""
        self.currentPosition = self.skip_delimiters(self.currentPosition)
        start = self.currentPosition
        self.currentPosition = self.scan_token(self.currentPosition)
        return self.text[start: self.currentPosition]

    def __iter__(self):
        """这个方法用于支持以迭代的方式返回符号"""
        while True:
            token = self.next_token()
            if(token==None):
                break
            yield token
```

使用 StringTokenizer 类：

```
it = iter(StringTokenizer("2#7#8#9#道","#"))
for x in range(5):   #取得 5 个标记
    print(next(it))
```

2.16.1 zip 函数

函数 zip() 遍历多个可迭代的对象，并聚合它们成为一个可迭代的 zip 对象。例如，把两个列表聚合成一个：

```
x = [1,2,3,4]
y = [7,8,3,2]
z = ['a','b','c','d']

for a,b in zip(x,y):
    print(a,b)
```

输出如下：

```
1 7
2 8
3 3
4 2
```

也可以聚合两个以上列表：

```
for a,b,c in zip(x,y,z):
    print(a,b,c)
```

输出如下：

```
1 7 a
2 8 b
3 3 c
4 2 d
```

可以使用计数器统计 zip 对象中元素出现的频次：

```
from collections import Counter

x = [1,2,3,4,4]
y = [7,8,3,2,2]

print(Counter(zip(x,y)))
```

输出如下：

```
Counter({(4, 2): 2, (1, 7): 1, (2, 8): 1, (3, 3): 1})
```

Counter.most_common()方法返回出现频次最高的 n 个元素，例如返回最常见的 3 个元素：

```
print(c.most_common(3))
```

输出如下：

```
[((4, 2), 2), ((1, 7), 1), ((2, 8), 1)]
```

2.16.2　itertools 模块

itertools 模块实现了许多迭代器构建块。函数 itertools.islice()用于从可迭代对象取切片。

```python
import itertools

def zigzag(period):
    while True:
        for n in range(period):
            yield n
        for n in range(period, 0, -1):
            yield n

it1 = zigzag(5)                          #创建一个无限长度的可迭代对象
it2 = itertools.islice(it1, 0, 20)       #取前20个元素
li = list(it2)
print(li)     #输出：[0, 1, 2, 3, 4, 5, 4, 3, 2, 1, 0, 1, 2, 3, 4, 5, 4, 3, 2, 1]
```

使用函数 itertools.count() 可创建自增 ID：

```
>>> import itertools
>>> id_iter = itertools.count()
>>> next(id_iter)
0
```

2.17 数据库

这里以数据库 MariaDB 为例，介绍 Python 的数据库客户端。

安装数据库 MariaDB：

```
#sudo apt-get install mariaDB-server mariaDB-client
```

以 MariaDB root 用户身份登录：

```
#sudo mysql -u root -p
```

输入密码后，将进入 MariaDB shell。

如果要启动 MariaDB，可以使用以下命令。

```
#sudo systemctl start mariadb
```

可以使用以下命令停止 MariaDB：

```
#sudo systemctl stop mariadb
```

安装 Python MySQL 客户端：

```
#python3 -m pip install PyMySQL
```

连接数据库：

```
import pymysql.cursors

#Connect to the database
```

```
connection = pymysql.connect(host='localhost',
                             user='user',
                             password='',
                             db='mysql',
                             charset='utf8mb4',
                             cursorclass=pymysql.cursors.DictCursor)
```

如果出现错误"pymysql.err.InternalError: (1698, "Access denied for user 'root'@'localhost'")", 则可以考虑使用 mysql_native_password 插件设置用户。

```
$ sudo mysql -u root
MariaDB [(none)]> USE mysql;

MariaDB [mysql]> UPDATE user SET plugin='mysql_native_password' WHERE User='root';
Query OK, 1 row affected (0.00 sec)
Rows matched: 1  Changed: 1  Warnings: 0

MariaDB [mysql]> FLUSH PRIVILEGES;
Query OK, 0 rows affected (0.00 sec)

MariaDB [mysql]> exit;
Bye
```

重新启动 MariaDB 服务:

```
#sudo systemctl restart mariadb
```

在以下示例中,获取 MariaDB 的版本信息。

```
import pymysql

con = pymysql.connect('localhost', 'root', '', 'mysql')

with con:

    cur = con.cursor()
    cur.execute("SELECT VERSION()")   #使用游标执行SQL语句
    version = cur.fetchone()

    print("Database version: {}".format(version[0]))
```

输出:

```
Database version: 10.1.38-MariaDB-0ubuntu0.18.04.2
```

插入数据和查找数据的例子:

```
import pymysql.cursors

#连接到数据库
```

```
connection = pymysql.connect(host='localhost',
                             user='root',
                             password='',
                             db='mysql',
                             charset='utf8mb4',
                             cursorclass=pymysql.cursors.DictCursor)

cursor=connection.cursor()

cursor.execute("CREATE TABLE IF NOT EXISTS results (dataset text, wer float)")
cursor.execute('INSERT INTO results(dataset, wer) VALUES(%s, %s)', ("LibriSpeech", 0.0583))

connection.commit()   #提交更新

cursor.close()

cursor=connection.cursor()

cursor.execute("select dataset, wer from results;")
rows = cursor.fetchall()

for row in rows:
    print(row["dataset"], row["wer"])
```

可以将数据库连接参数写入配置文件。Python 标准库中的 configparser 模块定义了用于读取和写入 Windows 操作系统使用的配置文件的功能。此类文件通常具有.ini 扩展名。INI 格式文件由 section 下的键/值对组成。sampleconfig.ini 文件内容示例如下：

```
[SectionName]
keyname1=value
;comment
keyname2=value
```

以下脚本读取并解析 sampleconfig.ini 文件：

```
import configparser
parser = configparser.ConfigParser()
parser.read('sampleconfig.ini')
for sect in parser.sections():
    print('Section:', sect)
    for k,v in parser.items(sect):
        print('  {} = {}'.format(k,v))
    print()
```

2.18 日志

Python 在标准库中随附了一个日志记录模块。该模块提供了一个灵活的框架从 Python 程序中发出日志消息。

日志记录模块的最重要的对象如下:
- Logger:实际的日志接口。
- Handler:处理日志语句并输出日志。
- Formatter:将输入数据格式化为字符串。
- Filter:允许过滤某些消息。

使用日志记录可以实现:
- 为消息设置不同的级别,例如 info()、debug()、error()、warning()。
- 添加消息的不同格式,即放置日期时间戳、文件名、进程、路径,以及更多 LogRecord 属性。
- 同时输出日志到文件、套接字等。

根据需要在 Python 程序中跟踪的消息或事件的严重性,可以使用不同的预定义级别。日志级别见表 2-4。

表 2-4 日志级别

级别	数值	什么时候使用
CRITICAL	50	严重错误,表明程序本身可能无法继续运行
ERROR	40	由于存在更严重的问题,该软件无法执行某些功能
WARNING	30	表示发生了意外情况,或者表示在不久的将来会出现一些问题(例如"磁盘空间不足")。该软件仍按预期运行
INFO	20	确认一切正常
DEBUG	10	详细信息,通常仅在诊断问题时才需要
NOTSET	0	创建日志时,级别设置为 NOTSET

日志的最低级别为 DEBUG,最高级别为 CRITICAL。因此,建议始终将日志级别定义为最小值,即 DEBUG,以便在需要时也可以使用较高的日志级别。

在如下的 Python 脚本中,导入了 logging 模块,但尚未定义任何级别。

```
#!/usr/bin/env python3

import logging

logging.debug('debug')
logging.info('info')
logging.warning('warning')
logging.error('error')
logging.critical('critical')
```

当执行这个脚本时，只得到了函数 warning()、error()和 critical()的输出。这是因为默认日志记录级别设置为 warning()，并且脚本将仅考虑所有具有与 warning()级别相等或更高数值的级别。

```
#python3 /tmp/logging_ex.py
WARNING:root:warning
ERROR:root:error
CRITICAL:root:critical
```

在上面例子的日志输出中得到了"root"。日志层次结构的根称为根日志。这就是函数 debug()、info()、warning()、error()和 critical()使用的日志，它们只是调用根日志的同名方法。根日志的名称在日志的输出中打印为"root"。

第 3 章

Flask 框架与微服务

使用 Flask 可以创建 Web 服务，使用 JavaScript 框架可以实现 Web 前端展示，使用模板引擎可以展现搜索结果。

本章首先介绍 Flask 及 Jinja 模板引擎，然后介绍通过 curl 命令行工具来测试 RESTful API。Flask 可以使用 JSON 作为响应报文格式，因此随后介绍 JSON 数据格式。

3.1 Flask 简介

Flask 是一个轻量级的 WSGI（Web 服务器网关接口）Web 应用程序框架。

在 Linux 下安装 Flask：

```
#pip3 install -U Flask
```

hello.py 文件中一个完整的 Flask 应用程序如下所示。

```
#cat ./hello.py
from flask import Flask

app = Flask(__name__)

@app.route('/')
def hello():
    return 'Hello, World!'
```

那么，这段代码做了什么？

首先导入了 Flask 类。此类的一个实例将是 WSGI 应用程序。

接下来创建这个类的一个实例。第一个参数是应用程序的模块或包的名称。__name__ 是一个适用于大多数情况的方便的快捷方式。这个参数是必需的，以便 Flask 知道在哪里寻找资源，例如模板和静态文件。

然后使用 route() 装饰器告诉 Flask 哪个 URL 应该触发函数。

函数返回要在用户浏览器中显示的消息。默认内容类型为 HTML，因此字符串中的 HTML 将由浏览器呈现。

启动服务：

```
#env FLASK_APP=hello.py flask run
 * Serving Flask app "hello.py"
 * Running on http://127.0.0.1:5000/ (Press CTRL+C to quit)
```

在 Windows 操作系统下，则需要用 set 命令先设置环境变量。

```
>set FLASK_APP=hello.py
```

启动应用：

```
>flask run
```

每当对应用程序代码进行任何更改时，都必须停止和启动服务器。但是，如果将环境变量 FLASK_ENV 设置为 development 然后运行服务器，每次对应用程序代码进行更改时，服务器都会自动重新加载代码。这还将启用调试器并在调试模式下运行应用程序。

如下命令启用调试模式：

```
>set FLASK_ENV=development
>flask run
```

除了使用环境变量，还可以在配置文件中设置 Flask 应用。安装 python-dotenv 包，在项目根文件夹中创建一个.flaskenv 文件并添加，例如：

```
FLASK_APP=hello.py
FLASK_ENV=development
```

保存文件。执行 flask run 命令。

如果运行服务器，会注意到该服务器只能从用户自己的计算机访问，而不能从网络中的任何其他计算机访问。这是默认设置，因为在调试模式下，应用程序的用户可以在自己的计算机上执行任意 Python 代码。

如果禁用了调试器或信任网络上的用户，则只需在命令行中添加--host=0.0.0.0 即可使服务器公开可用：

```
>flask run --host=0.0.0.0
```

这会告诉用户的操作系统侦听所有公共 IP。

3.2 模板引擎 Jinja

Jinja 是用于 Python 编程语言的 Web 模板引擎。本节介绍 Jinja 的基本使用及搜索结果输出。

3.2.1　Jinja 的基本使用

首先在交互式环境试用 Jinja 模板引擎。

```
>>> from jinja2 import Template
>>> t = Template("Hello {{ something }}!")
>>> t.render(something="World")
'Hello World!'
>>> t = Template("My favorite numbers: {% for n in range(1,10) %}{{n}} "  "{% endfor %}")
>>> t.render()
'My favorite numbers: 1 2 3 4 5 6 7 8 9 '
>>>
```

Jinja 中的分隔符包括：

- {%....%}：用于语句。
- {{....}}：是用于打印到模板输出的表达式。
- {#....#}：用于未包含在模板输出中的注释。
- #....##：用作行语句。

Jinja 允许使用方便的点表示法来访问 Python 词典中的数据。

```
from jinja2 import Template

person = { 'name': 'Person', 'age': 34 }

tm = Template("My name is {{ per.name }} and I am {{ per.age }}")
#注释掉的方法也是有效的
#tm = Template("My name is {{ per['name'] }} and I am {{ per['age'] }}")
msg = tm.render(per=person)

print(msg)
```

使用 FileSystemLoader 加载的文本文件。

```
from jinja2 import Environment, FileSystemLoader

persons = [
    {'name': 'Andrej', 'age': 34},
    {'name': 'Mark', 'age': 17},
    {'name': 'Thomas', 'age': 44},
    {'name': 'Lucy', 'age': 14},
    {'name': 'Robert', 'age': 23},
    {'name': 'Dragomir', 'age': 54}
]

file_loader = FileSystemLoader('templates')
env = Environment(loader=file_loader)
```

```python
#get the template with the get_template() method.
template = env.get_template('showpersons.txt')

output = template.render(persons=persons)
print(output)

with open("filename.txt", 'w',encoding='utf-8') as f:
    f.write(output)
```

templates/showpersons.txt 内容如下:

```
{% for person in persons -%}
    {{ person.name }} {{ person.age }}
{% endfor %}
```

显示搜索结果的代码如下:

```python
from jinja2 import Environment, FileSystemLoader

items = [
    {'url': 'http://lietu.com', 'title': '猎兔', 'body': 'content'},
    {'url': 'http://patfun.com', 'title': '健趣', 'body': 'content'}
]

file_loader = FileSystemLoader('templates')
env = Environment(loader=file_loader)

#get the template with the get_template() method.
template = env.get_template('searchPagination.txt')

output = template.render(resultItems=items,websiteTitle='title')
print(output)

with open("static/search.html", 'w', encoding='UTF-8') as f:
    f.write(output)
```

templates 目录下的搜索结果模板文件 searchPagination.txt 内容如下:

```html
<!DOCTYPE html>
<html>
<head>
    <meta http-equiv="Content-Type" content="text/html; charset=UTF-8">
    <meta name="viewport" content="width=device-width, initial-scale=1">
    <meta http-equiv="X-UA-Compatible" content="IE=edge">
    <title>{{websiteTitle}}</title>

    <link href="https://maxcdn.bootstrapcdn.com/bootstrap/4.3.1/css/bootstrap.min.css" rel="stylesheet">
```

```html
        <link href="../css/style.css" rel="stylesheet" type="text/css">
        <link href="../css/font-awesome.min.css" rel="stylesheet" type="text/css">
</head>
<body>

{% for result in resultItems -%}

        <ol class="list-unstyled col-md-8">
  <a href="{{ result.url }}">{{ result.title }}</a> <br/>
  {{ result.body }}
        </ol>

{% endfor %}

</body>
</html>
```

3.2.2 实现分页

实现分页的 PagerTag 类代码如下：

```python
import urllib

DEFAULT_ID = "pager"
OFFSET_PARAM = ".offset"

class PagerTag(object):

    def __init__(self, u: str, size: int, totalHits: int):
        self.url = u
        self.itemCount = totalHits   #条目总数
        self.maxPageItems = size
        self.offset = 0
        self.idOffsetParam = DEFAULT_ID + OFFSET_PARAM
        self.params = 0
        baseUri = None
        if (self.url != None) :
            baseUri = self.url;

        self.uri='';
        self.uri+=(baseUri);

    def set_offset(self, offset: int):
        self.offset = offset

    def add_param(self, name: str, value: str):
```

```python
        name = urllib.parse.quote(name)
        value = urllib.parse.quote(value)
        if self.params == 0:
            self.uri+=("?")
        else:
            self.uri+=("&amp;")
        self.uri+=(name);
        self.uri+=('=');
        self.uri+=(value);
        self.params += 1

    def has_next_page(self):
        return (self.itemCount > self.get_next_offset())

    def has_prev_page(self):
        return (self.offset > 0);

    def get_prev_url(self):
        return self.get_offset_url(self.get_prev_offset())

    def get_next_url(self):
        return self.get_offset_url(self.get_next_offset())

    def get_next_offset(self):
        return self.offset + self.maxPageItems

    def get_offset_url(self,pageOffset:int):
        uriLen = len(self.uri)
        self.uri+=("&amp;");
        self.uri+=(self.idOffsetParam);
        self.uri+=('=');
        self.uri+=str(pageOffset)
        offsetUrl = str(self.uri)
        self.uri = self.uri[0:uriLen]
        return offsetUrl
```

实现分页的模板 pagination.html 内容如下：

```
{% if (pager.has_prev_page()) %}
    <a href="{{pager.get_prev_url()}}" class="rnavLink">&#171;  上 一 页 </a>  
{% endif %}


{% if (pager.has_next_page()) %}
    <a href="{{pager.get_next_url()}}" class="rnavLink">下一页 &#187;</a>
{% endif %}
```

在代码中使用 PagerTag 类：

```
from jinja2 import Environment, FileSystemLoader
from PagerTag import PagerTag

file_loader = FileSystemLoader('templates')
env = Environment(loader=file_loader)


template = env.get_template('pagination.html')

pagination = PagerTag("./", 10,899)

keyWords = "二甲双胍";                    #查询词
pagination.add_param("query", keyWords)

output = template.render(pager=pagination)
print(output)

with open("static/searchPaginator.html", 'w', encoding='UTF-8') as f:
    f.write(output)
```

3.2.3 在 Flask 中使用 Jinja

app/templates/index.html 文件内容如下：

```
<!DOCTYPE html>
<html>
<p>Hello {{ username }}</p>
</body>
</html>
```

app/routes.py 文件内容如下：

```
from flask import Flask, render_template
app = Flask(__name__, template_folder='')
@app.route('/user/<username>')
def index(username):
    return render_template('index.html', username=username)
if __name__ == '__main__':
 app.run(debug=True)
```

这里使用 render_template()方法渲染 HTML。

3.3 测试 RESTful API 的 curl 命令

使用 curl 命令行工具可以发送 HTTP 网络请求来测试 RESTful API。

API 请求由 4 个不同部分组成：

- endpoint：客户端用于与服务器通信的 URL。
- HTTP 方法：告诉服务器客户端要执行什么操作。最常见的方法是 GET、POST、PUT、DELETE 和 PATCH。
- 标头：用于在服务器和客户端之间传递其他信息，如授权。
- body：发送到服务器的数据。

curl 命令的语法如下：

```
curl [options] [URL...]
```

在发出请求时，可以使用以下选项：

-X, --request：使用的 HTTP 方法。

-i, --include：包括响应头。

-d, --data：要发送的数据。

-H, --header：要发送的其他标头。

GET 方法从服务器请求特定资源。使用 curl 命令行工具发出 HTTP 请求时，GET 是默认方法。以下是向 JSONPlaceholder API 发出 GET 请求的示例：

```
#curl https://jsonplaceholder.typicode.com/posts
```

使用查询参数过滤结果：

```
#curl https://jsonplaceholder.typicode.com/posts?userId=1
```

POST 方法用于在服务器上创建资源。如果资源存在，则将其覆盖。以下命令将使用-d 选项指定的数据创建一个新帖子：

```
#curl -X POST -d "userId=5&title=Hello World&body=Post body." https://jsonplaceholder.typicode.com/posts
```

请求主体的类型使用 Content-Type 标头指定。默认情况下，如果不指定此标头，curl 使用 Content-Type: application/x-www-form-urlencoded。

```
#curl -X POST -H "Content-Type: application/json" \
    -d '{"userId": 5, "title": "Hello World", "body": "Post body."}' \
    https://jsonplaceholder.typicode.com/posts
```

要发送 JSON 格式的数据，则将主体类型设置为 application/json：

```
#curl -X POST -H "Content-Type: application/json" \
    -d '{"userId": 5, "title": "Hello World", "body": "Post body."}' \
    https://jsonplaceholder.typicode.com/posts
```

PUT 方法用于更新或替换服务器上的资源。它将指定资源的所有数据替换为请求数据。

```
#curl -X PUT -d "userId=5&title=Hello World&body=Post body." https://jsonplaceholder.typicode.com/posts/5
```

PATCH 方法用于对服务器上的资源进行部分更新。

```
#curl -X PATCH -d "title=Hello Universe" https://jsonplaceholder.typicode.
com/posts/5
```

DELETE 方法从服务器中删除指定的资源。

```
#curl -X DELETE https://jsonplaceholder.typicode.com/posts/5
```

如果 API 端点需要身份验证，则需要获取访问密钥；否则，API 服务器将使用"禁止访问"或"未经授权"响应消息来响应。获取访问密钥的过程取决于所使用的 API。获得访问令牌后，可以在标头中发送它：

```
#curl -H "Authorization: Basic <ACCESS_TOKEN>" http://www.example.com
```

3.4 JSON 数据格式

REST API 通常以 JSON 数据格式响应。JSON（JavaScript Object Notation）是一种轻量级的数据交换格式，很容易被阅读和编写，机器也很容易解析和生成。利用它可以传输由名称/值对和数组数据类型组成的数据对象。一些结构复杂的数据也可以采用 JSON 格式表示。例如，散列表中的值是数组。

JSON 的基本数据类型如下。

- 数字：有符号的十进制数字，可能包含小数部分，可能使用科学记数法，但不能包括非数字，如 NaN。该格式不区分整数和浮点数。
- 字符串：0 个或多个 Unicode 字符的序列。字符串用双引号分隔，并支持反斜杠转义语法。
- 布尔值：为 true 或 false 的任一值。
- 数组：0 个或多个值的有序列表，每个值可以是任何类型。数组使用方括号符号，元素以逗号分隔。
- 对象：名称/值对的无序集合，其中名称（也称为键）是字符串。由于对象旨在表示关联数组，推荐每个键在对象内是唯一的。对象用大括号分隔，并使用逗号分隔每一对，而在每一对中，冒号":"字符将键或名称与其值分隔开。
- null：一个空值，使用单词 null。

json 模块可以完成 JSON 的序列化和反序列化操作。

```
import json

cities = ["Bratislava",
          "Banská Bystrica",
          "Prešov",
          "Považská Bystrica",
          "Žilina",
```

```
            "Košice",
            "Ružomberok",
            "Zvolen",
            "Poprad",
            "新加坡",
            "新德里"]

print(json.dumps(cities))
```

通过 Response()构造方法返回 JSON 格式的数据的例子：

```
from flask import Flask, Response
import json

app = Flask(__name__)

cities = ["Bratislava",
          "Banská Bystrica",
          "Prešov",
          "Považská Bystrica",
          "Žilina",
          "Košice",
          "Ružomberok",
          "Zvolen",
          "Poprad",
          "新加坡",
          "新德里"]

@app.route('/_autocomplete', methods=['GET'])
def autocomplete():
    return Response(json.dumps(cities), mimetype='application/json')
```

3.5 构建 REST API

REST（Representational State Transfer，表征状态转移）是分布式超媒体系统的架构风格。HATEOAS（Hypermedia as the Engine of Application Statue，超媒体即应用状态引擎）是 REST 架构的主要约束。

使用函数 jsonify()实现 REST API：

```
#encoding: UTF-8
import json
from flask import Flask, jsonify
app = Flask(__name__)
app.config['JSON_AS_ASCII'] = False
```

```
@app.route('/')
def index():
    return jsonify({'url': 'http://www.patfun.com', 'title': '健趣网', 'body': 'content'})

app.run()
```

为了让 REST API 接收传递进来的参数，可以使用 request.args 获取从查询字符串解析出的内容：

```
from flask import request

@app.route('/')
def index():
    q = request.args.get('q')
    offset = request.args.get('offset')
```

REST API 完整的代码如下：

```
#encoding: UTF-8
import json
from flask import Flask, jsonify
from flask import request

app = Flask(__name__)
app.config['JSON_AS_ASCII'] = False
@app.route('/')
def index():
    q = request.args.get('q')
    offset = request.args.get('offset')
    data = [{
    "url":"http://patfun.com",
    "title":"健趣网",
    "body": "健趣网(www.patfun.com)是中国领先的健康垂直搜索,帮您一站找到最佳健康信息、健康产品及健康服务。"
    },
    {
    "url":"http://qq.com",
    "title":"QQ",
    "body": "腾讯网从2003年创立至今,已经成为集新闻信息、区域垂直生活服务、社会化媒体资讯和产品为一体的互联网媒体平台。"
    }
    ];
    return jsonify(q=q,data=data);

app.run()
```

集成 Elasticsearch 查询功能。

```
#encoding: UTF-8
import json
```

```python
from flask import Flask, jsonify
from flask import request
from elasticsearch import Elasticsearch

es = Elasticsearch()
app = Flask(__name__)
app.config['JSON_AS_ASCII'] = False
@app.route('/')
def index():
    q = request.args.get('q')
    offset = request.args.get('offset')

    res= es.search(index='sites',body={'query':{'match':{'body':q}}})
    print("match body Got %d Hits:" % res['hits']['total']['value'])
    print(res['hits']['hits'])

    jsons = res['hits']['hits']
    data = []
    for hits in jsons:
        itemDict = {}
        itemDict["url"] = hits["_source"]["url"]
        print(hits["_source"]["url"])
        itemDict["title"] = hits["_source"]["title"]
        itemDict["body"] = hits["_source"]["body"]
        data.append(itemDict)

    return jsonify(q=q,data=data);

app.run()
```

Node.js 可以调用 REST API。

使用 needle 模块调用 API 的例子如下：

```
>npm install needle
```

调用 API 的代码如下：

```
const needle = require('needle');
needle.get('https://jsonplaceholder.typicode.com/todos/1', {json: true}, (err, res) => {
    if (err) {
        return console.log(err);
    }
    let todo = res.body;
    console.log(todo.id);
    console.log(todo.title);
});
运行 testAPI.js
>node testAPI.js
```

3.6 Swagger 文档

Swagger 文档是一种用于可视化 RESTful Web 服务的规范。它代表 RESTful API，几乎可以与任何编程语言集成。这里通过示例演示生成 Swagger 文档。

Flask 应用 main.py 文件中的第一个端点代码如下：

```
#main.py
from flask import Flask
app = Flask(__name__)

@app.route('/basic_api/hello_world')
def hello_world():
    return 'Hello, World!'
```

首先从 Flask 导入 request 对象并将以下行添加到 main.py 文件中，以便为特定实体创建 CRUD 端点：

```
#main.py
from flask import Flask, request
...
@app.route('/basic_api/entities', methods=['GET', 'POST'])
def entities():
    if request.method == "GET":
        return {
            'message': 'This endpoint should return a list of entities',
            'method': request.method
        }
    if request.method == "POST":
        return {
            'message': 'This endpoint should create an entity',
            'method': request.method,
            'body': request.json
        }

@app.route('/basic_api/entities/<int:entity_id>', methods=['GET', 'PUT', 'DELETE'])
def entity(entity_id):
    if request.method == "GET":
        return {
            'id': entity_id,
            'message': 'This endpoint should return the entity {} details'.format(entity_id),
            'method': request.method
        }
    if request.method == "PUT":
```

```
            return {
                'id': entity_id,
                'message': 'This endpoint should update the entity {}'.format
(entity_id),
                'method': request.method,
                'body': request.json
            }
        if request.method == "DELETE":
            return {
                'id': entity_id,
                'message': 'This endpoint should delete the entity {}'.format
(entity_id),
                'method': request.method
            }
```

现在，对于每个创建的路由，允许使用不同的方法。以下端点可用：

- GET /entities：获取实体列表。
- POST / entities：创建实体。
- GET / entities/<entity_id>：获取实体信息。
- PUT / entities/<entity_id>：更新实体。
- DELETE / entities/<entity_id>：删除实体。

为了测试这些端点的工作方式，可以使用如下所示的 curl 请求：

```
curl -X POST -H "Content-Type: application/json" -d '{"ping": "pong"}'
http://localhost:5000/basic_api/entities
```

另一个在 API 开发中非常有用且常用的工具是 Postman。

为了开始插入蓝图模型，创建一个新文件夹 blueprints，在其中创建一个名为 basic_endpoints 的文件夹并在该文件夹中创建一个名为 __init__.py 的文件：

```
#blueprints/basic_endpoints/__ini__.py
from flask import Blueprint, request

blueprint = Blueprint('api', __name__, url_prefix='/basic_api')

@blueprint.route('/hello_world')
def hello_world():
    return {'message': 'Hello World!'}

@blueprint.route('/entities', methods=['GET', 'POST'])
def entities():
    if request.method == "GET":
        return {
            'message': 'This endpoint should return a list of entities',
            'method': request.method
```

```python
        }
    if request.method == "POST":
        return {
            'message': 'This endpoint should create an entity',
            'method': request.method,
            'body': request.json
        }

@blueprint.route('/entities/<int:entity_id>', methods=['GET', 'PUT', 'DELETE'])
def entity(entity_id):
    if request.method == "GET":
        return {
            'id': entity_id,
            'message': 'This endpoint should return the entity {} details'.format(entity_id),
            'method': request.method
        }
    if request.method == "PUT":
        return {
            'id': entity_id,
            'message': 'This endpoint should update the entity {}'.format(entity_id),
            'method': request.method,
            'body': request.json
        }
    if request.method == "DELETE":
        return {
            'id': entity_id,
            'message': 'This endpoint should delete the entity {}'.format(entity_id),
            'method': request.method
        }
```

现在 main.py 文件只需要加载创建的蓝图并将其注册到应用程序对象：

```python
#main.py
from flask import Flask
from blueprints.basic_endpoints import blueprint as basic_endpoints

app = Flask(__name__)
app.register_blueprint(basic_endpoints)

if __name__ == "__main__":
    app.run()
```

接下来演示一个模板端点。要做的第一件事是创建模板文件夹，并在该文件夹中插入 example.html 文件。

```html
<!-- templates/example.html -->
<html>
<a href="https://example.com">
    <div class="meme">
        <p class="top">{{top}}</p>
        <p class="bottom">{{bottom}}</p>
    </div>
</a>
</html>
```

请注意，{{ }}的模板中使用了两个变量。这是一种在模板中包含 Python 代码的特殊格式，允许呈现动态内容。现在模板已创建，可使用 Flask 加载它。通过在不同的文件中创建一个新的蓝图来演示这个例子：

```python
#blueprints/jinja_endpoint/__init__.py
from flask import Blueprint, request, render_template

blueprint = Blueprint('jinja_template', __name__, url_prefix='/jinja_template')

@blueprint.route('')
def get_template():
    top = request.args.get('top') if 'top' in request.args else ''
    bottom = request.args.get('bottom') if 'bottom' in request.args else ''

    return render_template('example.html', top=top, bottom=bottom)
```

不要忘记在 main.py 文件中注册这个蓝图：

```python
#main.py
...
from blueprints.jinja_endpoint import blueprint as jinja_template_blueprint
...
app.register_blueprint(jinja_template_blueprint)
...
```

在这里可以看到基于 URL 中发送的查询参数 top 和 bottom 变量的定义。

为刚刚创建的页面添加一些 CSS。创建包含文件 example.css 的静态文件夹：

```css
/* static/example.css */
* {
    box-sizing: border-box;
}

.meme {
    margin: auto;
    width: 450px;
    height: 450px;
    background-image: url('https://www.imaginarycloud.com/blog/content/images/
```

```css
2021/03/business_cat.jpg');
    background-size: 100%;
    text-align: center;
    position: relative;
}

p {
    position: absolute;
    left: 0;
    right: 0;
    margin: 15px 0;
    padding: 0 5px;
    font-family: impact;
    font-size: 2.5em;
    text-transform: uppercase;
    color: white;
    letter-spacing: 1px;
    text-shadow:2px 2px 0 #000,
    -2px -2px 0 #000,
    2px -2px 0 #000,
    -2px 2px 0 #000,
    0px 2px 0 #000,
    2px 0px 0 #000,
    0px -2px 0 #000,
    -2px 0px 0 #000,
    2px 2px 5px #000;
}

.bottom {
    bottom: 0;
}

.top {
    top: 0;
}
```

之后，通过将这个 head 标签添加到 HTML 文件中，为刚刚创建的 CSS 文件添加相应的引用：

```html
<!-- templates/example.html -->
<html>
<head>
    <link rel="stylesheet" href="{{ url_for('static', filename='example.css') }}">
</head>
...
```

打开如下网址可以浏览模板输出的页面：

http://localhost:5000/jinja_template?top=cancel%20the%20REST%20API%20creation&bottom=I%20have%20to%20watch%20this%20bird

接下来，为端点创建更好的项目结构和文档。通过使用 Flask 扩展 Flask-RESTlus，可以自动生成具有更好结构的文档。

对于记录的端点，可以创建一个文件夹来标识正在处理的实体。随着项目变得越来越大且越来越复杂，这将有助于在项目中导航。documented_endpoints 中的每个文件夹都像具有与该实体相关的功能/端点的模块一样工作。

下面将展示所有已使用新结构创建的端点。端点将没有逻辑，但可以让读者了解其创建过程中涉及的步骤。

首先，应该考虑具有新结构的 hello_world 端点：

```python
#blueprints/documented_endpoints/hello_world/__init__.py
from flask import request
from flask_restplus import Namespace, Resource, fields

namespace = Namespace('hello_world', 'Hello World related endpoints')

hello_world_model = namespace.model('HelloWorld', {
    'message': fields.String(
        readonly=True,
        description='Hello world message'
    )
})

hello_world_example = {'message': 'Hello World!'}

@namespace.route('')
class HelloWorld(Resource):

    @namespace.marshal_list_with(hello_world_model)
    @namespace.response(500, 'Internal Server error')
    def get(self):
        '''Hello world message endpoint'''

        return hello_world_example
```

现在分解上面的代码。读者可能会注意到的第一件事是 Namespace 和 Resource 类的使用。Namespace 直接链接到特定实体，这意味着所有 hello_world 端点都将链接到相应的命名空间。这将在 swagger 文档中生成 hello_world 部分。

为了在这个命名空间中创建路由，从 Resource 类继承的不同类与各自的命名空间路由装饰器一起声明。在创建的 HelloWorld 类中，声明它考虑的方法。在我们的例子中，只提供了 GET，从而产生了一个可用于 docummented_endpoints/hello_world 端点的 GET 方法。

与创建的资源和相应方法相关的注释将在最终文档中生成描述。端点可能返回的可能错误也应按照示例中的说明进行指定，并且还将产生进一步的端点文档。flask-restplus 扩展支持更多功能，例如解析器和错误处理。

现在展示将所创建的命名空间链接到蓝图（blueprints/documented_endpoints/__init__.py）的代码，然后将蓝图链接到应用程序（main.py）：

```
#blueprints/documented_endpoints/__init__.py
from flask import Blueprint
from flask-restplus import Api
from blueprints.documented_endpoints.hello_world import namespace as hello_world_ns

blueprint = Blueprint('documented_api', __name__, url_prefix='/documented_api')

api_extension = Api(
    blueprint,
    title='Flask RESTplus Demo',
    version='1.0',
    description='Application tutorial to demonstrate Flask RESTplus extension\
        for better project structure and auto generated documentation',
    doc='/doc'
)

api_extension.add_namespace(hello_world_ns)

#main.py
from flask import Flask
from blueprints.basic_endpoints import blueprint as basic_endpoint
from blueprints.jinja_endpoint import blueprint as jinja_template_blueprint
from blueprints.documented_endpoints import blueprint as documented_endpoint

app = Flask(__name__)
app.config['RESTPLUS_MASK_SWAGGER'] = False

app.register_blueprint(basic_endpoint)
app.register_blueprint(jinja_template_blueprint)
app.register_blueprint(documented_endpoint)

if __name__ == "__main__":
    app.run()
```

在这里，创建了一个来自 fask-restplus 模块的 API 对象并链接到文档化的 API 蓝图。蓝图和所创建的 hello_world 命名空间之间的链接是通过使用文档对象的 add_namespace 方法完成的。最后，就像之前所有的蓝图一样，只需要将创建的蓝图注册到 app 对象中。访问链接 http://localhost:5000/documented_api/doc 可以看到 Flask RESTplus 自动生成的页面。

现在展示实体 REST API 和 Jinja 模板页面文档的代码，以便任何人都可以轻松检查在处理的所有端点。这是通过添加以下文件来完成的：

```python
#blueprints/documented_endpoints/entities/__init__.py
from flask import request
from flask_restplus import Namespace, Resource, fields
from http import HTTPStatus


namespace = Namespace('entities', 'Entities fake endpoints')

entity_model = namespace.model('Entity', {
    'id': fields.Integer(
        readonly=True,
        description='Entity identifier'
    ),
    'name': fields.String(
        required=True,
        description='Entity name'
    )
})

entity_list_model = namespace.model('EntityList', {
    'entities': fields.Nested(
        entity_model,
        description='List of entities',
        as_list=True
    ),
    'total_records': fields.Integer(
        description='Total number of entities',
    ),
})

entity_example = {'id': 1, 'name': 'Entity name'}

@namespace.route('')
class entities(Resource):
    '''Get entities list and create new entities'''

    @namespace.response(500, 'Internal Server error')
    @namespace.marshal_list_with(entity_list_model)
    def get(self):
        '''List with all the entities'''
        entity_list = [entity_example]

        return {
```

```python
            'entities': entity_list,
            'total_records': len(entity_list)
        }

    @namespace.response(400, 'Entity with the given name already exists')
    @namespace.response(500, 'Internal Server error')
    @namespace.expect(entity_model)
    @namespace.marshal_with(entity_model, code=HTTPStatus.CREATED)
    def post(self):
        '''Create a new entity'''

        if request.json['name'] == 'Entity name':
            namespace.abort(400, 'Entity with the given name already exists')

        return entity_example, 201

@namespace.route('/<int:entity_id>')
class entity(Resource):
    '''Read, update and delete a specific entity'''

    @namespace.response(404, 'Entity not found')
    @namespace.response(500, 'Internal Server error')
    @namespace.marshal_with(entity_model)
    def get(self, entity_id):
        '''Get entity_example information'''

        return entity_example

    @namespace.response(400, 'Entity with the given name already exists')
    @namespace.response(404, 'Entity not found')
    @namespace.response(500, 'Internal Server error')
    @namespace.expect(entity_model, validate=True)
    @namespace.marshal_with(entity_model)
    def put(self, entity_id):
        '''Update entity information'''

        if request.json['name'] == 'Entity name':
            namespace.abort(400, 'Entity with the given name already exists')

        return entity_example

    @namespace.response(204, 'Request Success (No Content)')
    @namespace.response(404, 'Entity not found')
    @namespace.response(500, 'Internal Server error')
    def delete(self, entity_id):
        '''Delete a specific entity'''
```

```python
    return '', 204
#blueprints/documented_endpoints/jinja_template/__init__.py
from flask import request, render_template, make_response
from flask_restplus import Namespace, Resource, reqparse

namespace = Namespace(
    'jinja_template',
    'Jinja Template page. Note that this is a html page and not a REST API endpoint')

parser = reqparse.RequestParser()
parser.add_argument('top', type=str, help='Top text')
parser.add_argument('bottom', type=str, help='Bottom text')


@namespace.route('')
class JinjaTemplate(Resource):

    @namespace.response(200, 'Render jinja template')
    @namespace.response(500, 'Internal Server error')
    @namespace.expect(parser)
    def get(self):
        '''Render jinja template page'''

        top = request.args.get('top') if 'top' in request.args else ''
        bottom = request.args.get('bottom') if 'bottom' in request.args else ''

        return make_response(render_template('example.html', top=top, bottom=bottom), 200)
```

尽管这些文件看起来很吓人，但其逻辑与 hello_world 端点呈现的逻辑相同。主要区别在于使用模型来生成文档并验证以 POST 和 PUT 方法发送的请求正文。解析器还用于查询参数文档，也可用于验证。这对于不通过正文发送数据的场景很有用，例如查询参数或 FormData。

最后，不要忘记添加以下几行以将创建的命名空间链接到正在创建的蓝图。

```python
#blueprints/documented_endpoints/__init__.py
...
from blueprints.documented_endpoints.entities import namespace as entities_ns
from blueprints.documented_endpoints.jinja_template import namespace as jinja_template_ns
...
api_extension.add_namespace(entities_ns)
api_extension.add_namespace(jinja_template_ns)
```

3.7 使用 Fetch API

创建一个简单的 Flask 应用程序，该应用程序可以使用 Fetch API 将数据提供给 Web 界面。

```
>flask run
```

构建包含一个空模板文件夹和一个 app.py 文件的存储库。

app.py 文件包含创建 Web 界面所需的数据。

```
########imports ##########
from flask import Flask, jsonify, request, render_template
app = Flask(__name__)
#############################
#Additional code goes here #
#############################
#########run app #########
app.run(debug=True)
```

定义 Flask 应用后，还需要创建一个模板网页。可以通过将文件 index.html 放在模板目录中来完成。

```
<body>
<h1> Python Fetch Example</h1>
<p id='embed'>{{embed}}</p>
<p id='mylog'/>
<body>
```

{{embed}}将由下面的代码段中的 embed_example 字符串替换。

```
@app.route('/')
def home_page():
    example_embed='This string is from python'
    return render_template('index.html', embed=example_embed)
```

在 app.py 中，可以为 GET 请求创建一个 URL。以下代码定义了调用 URL 时的响应。

```
@app.route('/test', methods=['GET', 'POST'])
def testfn():
    #GET request
    if request.method == 'GET':
        message = {'greeting':'Hello from Flask!'}
        return jsonify(message)  #serialize and use JSON headers
    #POST request
    if request.method == 'POST':
        print(request.get_json())  #parse as JSON
        return 'Sucesss', 200
```

在 GET 请求后，定义一个包含 greeting 元素的字典并将其序列化。接下来，将其发布给调用 JavaScript 程序。

可以访问 https://127.0.0.1:5000/test 得到以下结果：

```
{
    "greeting": "Hello from Flask!"
}
```

现在已经设置了服务器端，然后就可以使用 fetch 命令从中检索数据。为此，可以使用以下 fetch 承诺：

```
fetch('/test')
    .then(function (response) {
        return response.json();
    }).then(function (text) {
        console.log('GET response:');
        console.log(text.greeting);
    });
```

在这里，在/test 上运行 GET 请求，该请求将返回的 JSON 字符串转换为对象，然后将 greeting 元素打印到 Web 控制台。通常，JavaScript 代码应嵌套在 HTML 文档中的<script>标记之间。

现在，已有一个有效的示例，可以对其进行扩展以包括实际数据。实际上，这可能涉及访问数据库和解密某些信息或过滤表。

创建一个数据数组：

```
########Example data, in sets of 3 ############
data = list(range(1,300,3))
print (data)
```

在 Flask 应用程序中，可以向 GET 请求添加可选参数——数组索引。这里的可选参数是通过页面/getdata/<index_no>中的附加页面扩展名指定的。然后将此参数传递到页面函数中，并在 return 命令中处理（data[int(index_no)]）。

```
########Data fetch ############
@app.route('/getdata/<index_no>', methods=['GET','POST'])
def data_get(index_no):

    if request.method == 'POST': #POST request
        print(request.get_text())  #parse as text
        return 'OK', 200

    else: #GET request
        return 't_in = %s ; result: %s ;'%(index_no, data[int(index_no)])
```

除了更改 GET URL 以包括感兴趣的数据元素的索引外，获取请求保持不变。这是在 index.html 的 JS 脚本中完成的。

```
var index = 33;
fetch(`/getdata/${index}`)
    .then(function (response) {
        return response.text();
    }).then(function (text) {
        console.log('GET response text:');
        console.log(text);
    });
```

这次是返回一个字符串并解析它,而不是返回一个对象。

3.8 发布 Flask 到 Nginx

基本的 Flask Nginx 配置如下所示:

```
location = /yourapplication { rewrite ^ /yourapplication/; }
location /yourapplication { try_files $uri @yourapplication; }
location @yourapplication {
  include uwsgi_params;
  uwsgi_pass unix:/tmp/yourapplication.sock;
}
```

此配置将应用程序绑定到/yourapplication。如果要在 URL 根目录中添加它,则要简单一些:

```
location / { try_files $uri @yourapplication; }
location @yourapplication {
    include uwsgi_params;
    uwsgi_pass unix:/tmp/yourapplication.sock;
}
```

3.9 启用 HTTPS

HTTPS(HyperText Transfer Protocol over Secure Socket Layer,超文本传输安全协议)在 HTTP 的基础上通过传输加密和身份认证保证了传输过程的安全性。SSL(Secure Socket Layer,安全套接层)是基于 HTTPS 的一个协议加密层。

为了让网站支持 HTTPS,可以从腾讯云这样的网站申请 SSL 证书,得到 lietu.com.jks 和 keystorePass.txt 这样的文件。

然后修改 Nginx 配置文件 nginx.conf,增加对 HTTPS 的支持。

```
http {
    include       mime.types;
```

```
    default_type  application/octet-stream;
    sendfile        on;
    keepalive_timeout  65;
server {
    #监听443端口
    listen 443;
    #你的域名
    server_name lietu.com;
    ssl on;
    #SSL证书的pem文件路径
    ssl_certificate    /root/card/lietu.com.pem;
    #SSL证书的key文件路径
    ssl_certificate_key /root/card/lietu.com.key;
    location / {
     proxy_pass http://公网地址:项目端口号;
    }
  }
    server {
        listen 80;
        server_name lietu.com;
        #将请求转换成HTTPS
        rewrite ^(.*)$ https://$host$1 permanent;
    }
}
```

第 4 章

Flask 源代码分析

Pocoo 是一个松散组建的开源开发人员团队，致力于一些非常受欢迎的 Python 项目。微框架 Flask 基于 Pocoo 项目 Werkzeug 和 Jinja2。

4.1 Werkzeug 库

Werkzeug 是一个全面的 WSGI Web 应用程序库。它最初是 WSGI 应用程序的各种实用程序的简单集合，现已成为最高级的 WSGI 实用程序库之一。

Werkzeug 库包括：

- 交互式调试器，允许使用交互式解释器检查浏览器中的堆栈跟踪和源代码，以查看堆栈中的任何帧。
- 功能齐全的请求对象，具有可与标头、查询参数、表单数据、文件和 Cookie 交互的对象。
- 可以包装其他 WSGI 应用程序并处理流数据的响应对象。
- 一种路由系统，用于将 URL 匹配到端点并为端点生成 URL，并且具有可扩展的系统来捕获 URL 中的变量。
- HTTP 实用程序，用于处理实体标签、缓存控制、日期、用户代理、Cookie 和文件等。
- 在本地开发应用程序时使用的线程化 WSGI 服务器。
- 一个测试客户端，用于在测试过程中模拟 HTTP 请求，而无须运行服务器。

Flask 使用了 Werkzeug 路由系统，该系统旨在根据复杂度自动排序路由。这意味着用户可以任意顺序声明路由，它们仍将按预期工作。

Werkzeug 路由系统的另一个设计决策是，Werkzeug 中的路由会尝试确保 URL 唯一。如果路由不明确，Werkzeug 将自动重定向到规范 URL。

4.1.1 WSGI 简介

WSGI 是类似于 Servlet 规范的通用接口规范。与 Servlet 相似，只要编写的程序符合 WSGI 规范，它就可以在支持 WSGI 规范的 Web 服务器中运行，就像符合 Servlet 规范的应用程序可以在 Tomcat 和 Jetty 中运行一样。例如：

```
from wsgiref import simple_server

def application(environ, start_response):
    start_response('200 OK', [('Content-Type', 'text/plain')])
    return [b'Hello World!']

http_server = simple_server.make_server('0.0.0.0', 5000, application)
http_server.serve_forever()
```

请注意，在返回数据时，不要直接返回字符串，而是要返回字节。

可以看到 WSGI 程序的定义仅需要实现一个应用程序。非常简单的 3 行代码即可实现 HTTP 请求的处理。其中，environ 参数是一个 dict，包含系统的环境变量和 HTTP 请求的相关参数。

再来看一下 start_response 函数，第一个参数是带有描述的状态码；第二个参数是带有响应头的列表；返回值代表响应主体。

4.1.2 Werkzeug 演示

使用 Werkzeug 实现一个简单的 Web 服务器。

```
from wsgiref import simple_server

from werkzeug.wrappers import Request, Response

def application(environ, start_response):
    request = Request(environ)
    text = 'Hello %s!' % request.args.get('name', 'World')
    response = Response(text, mimetype='text/plain')
    return response(environ, start_response)

http_server = simple_server.make_server('0.0.0.0', 5000, application)
http_server.serve_forever()
```

在浏览器输入 http://localhost:5000/ 就可以看到结果。

如果自己编写 WSGI 程序，则需要自己解析环境并自己处理返回值。使用 Werkzeug 就可以通过库提供的 Request 和 Response 类简化开发。

Request 类继承了许多类，可以看到存在 Accept、ETAG、CORS 和其他相关的标头解析。

```python
class Request(
    BaseRequest,
    AcceptMixin,
    ETagRequestMixin,
    UserAgentMixin,
    AuthorizationMixin,
    CORSRequestMixin,
    CommonRequestDescriptorsMixin,
):
```

BaseRequest 的构造方法是：

```python
def __init__(self, environ, populate_request=True, shallow=False):
    self.environ = environ
    if populate_request and not shallow:
        self.environ["werkzeug.request"] = self
    self.shallow = shallow
```

由于 Request 类的方法和属性很多，因此这里分析一些常见的方法。request.query_string 属性实现如下：

```python
query_string = environ_property(
    "QUERY_STRING",
    "",
    read_only=True,
    load_func=lambda x: x.encode("latin1"),
    doc="The URL parameters as raw bytes.",
)
```

environ_property 是一个实现查找方法的类。

request.data 属性是为了获取请求正文。

request.args 的赋值方式是获取 QUERY_STRING，然后通过"&"切分，再使用"="切出一个键/值对，前一个用作键，后一个用作值。应该注意的是，这里使用了 MultiDict，目的是为同一个键存储多个值。

```python
def url_decode(
    s,
    charset="utf-8",
    decode_keys=None,
    include_empty=True,
    errors="replace",
    separator="&",
    cls=None,
):
    if cls is None:
        from .datastructures import MultiDict

        cls = MultiDict
```

```
        if isinstance(s, str) and not isinstance(separator, str):
            separator = separator.decode(charset or "ascii")
        elif isinstance(s, bytes) and not isinstance(separator, bytes):
            separator = separator.encode(charset or "ascii")
        return cls(_url_decode_impl(s.split(separator), charset, include_empty,
errors))
```

request.path()方法实现如下：

```
def path(self):
    raw_path = _wsgi_decoding_dance(
        self.environ.get("PATH_INFO") or "", self.charset, self.encoding_errors
    )
    return "/" + raw_path.lstrip("/")
```

Response 类有两个核心功能：一个是通过封装构造返回值；另一个是返回符合 WSGI 规范的函数。

```
#Response's init function
def __init__(
    self,
    response=None,
    status=None,
    headers=None,
    mimetype=None,
    content_type=None,
    direct_passthrough=False,
)

#Response call function
def __call__(self, environ, start_response):
    app_iter, status, headers = self.get_wsgi_response(environ)
    start_response(status, headers)
    return app_iter
```

以一个 Demo 为例，看一下 Map、Rule 类的用法。

```
from wsgiref import simple_server

from werkzeug.routing import Map, Rule, HTTPException
from werkzeug.wrappers import Response, Request

url_map = Map([
    Rule('/test1', endpoint='test1'),
    Rule('/test2', endpoint='test2'),
])

def test1(request, **args):
    return Response('test1')
```

```python
def test2(request, **args):
    return Response('test2')

views = {'test1': test1, 'test2': test2}

def application(environ, start_response):
    request = Request(environ)
    try:
        return url_map.bind_to_environ(environ) \
            .dispatch(
                lambda endpoint, args: views[endpoint](request, **args)
            )(environ, start_response)
    except HTTPException as e:
        return e(environ, start_response)

http_server = simple_server.make_server('0.0.0.0', 5000, application)
http_server.serve_forever()
```

每个规则都代表一个 URL 匹配模式，第一个参数字符串可以放<converter(arguments): name>，例如/all/page/<int:page>。端点可以放置字符串、函数等，这意味着如果匹配了相应的路径，则将返回端点的值。因为大多数应用程序将至少具有一个接口，所以规则存在的意义是可以定义从路径到特定处理函数（或由字符串表示的函数）的映射，从而简化了多个接口的开发。

Map 可以存储多个规则。url_map 调用 bind_to_environ()方法后，返回一个 MapAdapter 对象。

收到请求后，首先通过函数 bind_to_environ()获得 MapAdapter。

```python
def bind_to_environ(self, environ, server_name=None, subdomain=None):
    environ = _get_environ(environ)
    wsgi_server_name = get_host(environ).lower()
    scheme = environ["wsgi.url_scheme"]

    #Existence reduction

    def _get_wsgi_string(name):
        val = environ.get(name)
        if val is not None:
            return _wsgi_decoding_dance(val, self.charset)

    script_name = _get_wsgi_string("SCRIPT_NAME")
    path_info = _get_wsgi_string("PATH_INFO")
    query_args = _get_wsgi_string("QUERY_STRING")
    return Map.bind(
        self,
```

```
        server_name,
        script_name,
        subdomain,
        scheme,
        environ["REQUEST_METHOD"],
        path_info,
        query_args=query_args,
    )
```

函数 bind_to_environ()的主要逻辑可以理解为通过 environ 获取一些参数，然后调用 bind()方法，最后通过这些参数创建 MapAdapter 对象。

```
def bind(
    self,
    server_name,
    script_name=None,
    subdomain=None,
    url_scheme="http",
    default_method="GET",
    path_info=None,
    query_args=None,
):
    #Existence reduction
    return MapAdapter(
        self,
        server_name,
        script_name,
        subdomain,
        url_scheme,
        path_info,
        default_method,
        query_args,
    )
```

4.2 源代码分析

从最基本的 Flask 使用开始，代码如下：

```
from flask import Flask
app = Flask(__name__)
@app.route('/')
def hello():
        return "hello lazy programming"
if __name__ == '__main__':
    app.run()
```

在代码中，Flask(__name__)将实例化 Flask 类，并且相应的__init__()方法如下所示：

```
#flask/app.py
classs Flask(_PackageBoundObject):
    def __init__(...):
            #For each request, create a processing channel.
        self.config = self.make_config()
        self.view_functions = {}
        self.error_handler_spec = {}
        self.before_request_funcs = {}
        self.before_first_request_funcs = []
        self.after_request_funcs = {}
        self.teardown_request_funcs = {}
        self.teardown_appcontext_funcs = []
        self.url_value_preprocessors = {}
        self.url_default_functions = {}
        self.url_map = Map()
        self.blueprints = {}
        self._blueprint_order = []
        self.extensions = {}
```

实例化 Flask 之后，请使用@app.route('/')装饰器方法将 hello()方法映射到路由。相关代码如下：

```
#flask/app.py/Flask
def route(self, rule, **options):
    def decorator(f):
        endpoint = options.pop('endpoint', None)
        self.add_url_rule(rule, endpoint, f, **options)
        return f
    return decorator
```

route()方法是一个简单的装饰器，具体的处理逻辑在 add_url_rule()方法中。

```
#flask/app.py/Flask
@setupmethod
def add_url_rule(self, rule, endpoint=None, view_func=None,
                 provide_automatic_options=None, **options):
    #... omit other code details
    rule = self.url_rule_class(rule, methods=methods, **options)
    rule.provide_automatic_options = provide_automatic_options
        #Add routing rule to url_map
    self.url_map.add(rule)
    if view_func is not None:
        old_func = self.view_functions.get(endpoint)
            #The endpoint of each method must be different
        if old_func is not None and old_func != view_func:
            raise AssertionError('View function mapping is overwriting an '
```

```
                    'existing endpoint function: %s' % endpoint)
              #Map the endpoint corresponding to the rule to view_func through
the view_functions dictionary
        self.view_functions[endpoint] = view_func
```

从 add_url_rule()方法可以看到，@app.route('/')主要功能是将路径保存到 url 映射，将装饰方法保存到 viewfunctions。应当注意，每个方法的端点必须不同，否则将引发 AssertionError。

最后，调用 app.run()方法来运行 Flask 应用程序，相应的代码如下：

```
#flask/app.py/Flask
def run(self, host=None, port=None, debug=None,
        load_dotenv=True, **options):
    #... omitted
    from werkzeug.serving import run_simple
    try:
        #Use
        run_simple(host, port, self, **options)
    finally:
        self._got_first_request = False
```

run()方法进一步调用 werkzeug.serving 下的 run_simple()方法以启动 Web 服务。这里的 self 是 Flask()的应用程序。

更深入的调用链是：runsimple()->makeserver()->BaseWSGIServer()->WSGIRequestHandler。

从名称中可以得知 WSGIRequestHandler 类的功能。它主要用于处理满足 WSGI 协议的请求。WSGIRequestHandler 类中代码的 execute()方法部分如下：

```
#werkzeug/serving.py/WSGIRequestHandler
def execute(app):
    application_iter = app(environ, start_response)
        #Omit other codes
```

简而言之，app.run()将启动一个符合 WSGI 协议的 Web 服务，它将侦听指定的端口，将 HTTP 请求解析为 WSGI 格式的数据，然后将 environ 和 start_response 传递给 Flask()实例对象。

作为方法调用该类对象，需要查看__call__()方法中的代码。

```
#flask/app.py/Flask
def __call__(self, environ, start_response):
    return self.wsgi_app(environ, start_response)
def wsgi_app(self, environ, start_response):
    #Request context
    ctx = self.request_context(environ)
    error = None
```

```
        try:
            try:
                ctx.push()
                    #The correct request processing path will find the corresponding processing function through the route
                    #Find the corresponding processing function through routing
                response = self.full_dispatch_request()
            except Exception as e:
                    #Error handling
                error = e
                response = self.handle_exception(e)
            except:
                error = sys.exc_info()[1]
                raise
            return response(environ, start_response)
        finally:
            if self.should_ignore_error(error):
                error = None
                #No matter whether an exception occurs or not, you need to error pop out from the request context
            ctx.auto_pop(error)
```

主要逻辑位于 wsgi_app()方法。在开始时处理请求上下文，然后调用 full_dispatch_request()方法，并获取返回值。如果请求路由不存在，它将执行错误处理并返回 500 错误。

full_dispatch_request()方法代码如下：

```
#flask/app.py/Flask
def full_dispatch_request(self):
    self.try_trigger_before_first_request_functions()
    try:
        request_started.send(self)
        rv = self.preprocess_request()
        if rv is None:
            #Call the processing function corresponding to the route and return the result of the processing function.
            rv = self.dispatch_request()
    except Exception as e:
        rv = self.handle_user_exception(e)
    return self.finalize_request(rv)
```

full_dispatch_request()方法中最关键的逻辑在于 dispatch_request()方法，该方法将调用相应路由的处理函数并获得该函数的结果。

最后应注意，app.run()方法仅在开发环境中使用。通过上面的分析，可知 app.run()的背后是使用 Werkzeug 构建简单的 Web 服务，但是该 Web 服务并不可靠，生产环境通常使用 uWSGI 启动 Flask 应用程序。

第 5 章

SQLAlchemy 操作数据库

SQLAlchemy 是 Python SQL 工具箱和对象关系映射器。它为应用程序开发人员提供了 SQL 的全部功能和灵活性。从一开始，它就试图提供一个端到端系统，以使用 Python 数据库 API（DBAPI）进行数据库交互来处理关系数据库。

5.1 使用 SQLAlchemy

安装 SQLAlchemy：

```
>pip install SQLAlchemy
```

升级版本：

```
>pip install --upgrade SQLAlchemy
```

安装指定版本：

```
>pip install SQLAlchemy==1.4.0b2
```

查看版本：

```
>>> import sqlalchemy
>>> sqlalchemy.__version__
'1.4.0b2'
```

5.2 SQL 表达式语言

SQL 表达式语言通过隐藏 SQL 语言并且不混合使用 Python 代码和 SQL 代码来提高代码的可维护性。

这里使用内存中的 SQLite 数据库来演示 SQL 表达式语言的使用。这是测试 SQLAlchemy 的简便方法，因为无须在任何地方定义实际的数据库。使用函数 create_engine()

连接数据库:

```
from sqlalchemy import create_engine
engine = create_engine('sqlite:///:memory:')
```

5.2.1 定义和创建表

创建表的函数使用示例如下:

```
metadata = MetaData()

user = Table('user', metadata,
    Column('user_id', Integer, primary_key=True),
    Column('user_name', String(16), nullable=False),
    Column('email_address', String(60), key='email'),
    Column('nickname', String(50), nullable=False)
)

user_prefs = Table('user_prefs', metadata,
    Column('pref_id', Integer, primary_key=True),
    Column('user_id', Integer, ForeignKey("user.user_id"), nullable=False),
    Column('pref_name', String(40), nullable=False),
    Column('pref_value', String(100))
)

addresses = Table('addresses', metadata,
  Column('id', Integer, primary_key=True),
  Column('user_id', None, ForeignKey('users.id')),
  Column('email_address', String, nullable=False)
 )

metadata.create_all(engine)
```

create_all()方法通常在表定义本身内联的表之间创建外键约束,因此它也按其依赖关系的顺序生成表。有一些选项可以更改此行为,使用 ALTER TABLE 来代替生成表。

使用 drop_all()方法可以通过类似的方法删除所有表。此方法与 create_all()方法的功能完全相反——首先检查每个表的存在,然后按照相反的依赖关系顺序删除表。

通过 Table 实例的 create()和 drop()方法可以创建和删除单个表。这些方法在默认情况下发出 CREATE 或 DROP 语句,而不管表的存在情况如何。

```
engine = create_engine('sqlite:///:memory:')

meta = MetaData()

employees = Table('employees', meta,
    Column('employee_id', Integer, primary_key=True),
```

```
    Column('employee_name', String(60), nullable=False, key='name'),
    Column('employee_dept', Integer, ForeignKey("departments.department_id"))
)
employees.create(engine)
```

drop()方法：

```
employees.drop(engine)
```

要启用"首先检查现有表"逻辑，请将 checkfirst = True 参数添加到 create()或 drop()方法：

```
employees.create(engine, checkfirst=True)
employees.drop(engine, checkfirst=True)
```

5.2.2 模式

大多数数据库都支持多个"模式"的概念——用来引用表和其他结构的可选择集合的命名空间。

使用 Table.schema 参数可以将"模式"名称直接与表关联。使用 Table 对象的 Table.schema 参数的示例如下所示：

```
metadata = MetaData()

financial_info = Table(
    'financial_info',
    metadata,
    Column('id', Integer, primary_key=True),
    Column('value', String(100), nullable=False),
    schema='remote_banks'
)
```

使用此表呈现的 SQL（例如下面的 SELECT 语句）将使用 remote_banks 模式名称显式限定表名 financial_info：

```
>>> print(select(financial_info))
SELECT remote_banks.financial_info.id, remote_banks.financial_info.value
FROM remote_banks.financial_info
```

当使用显式模式名称声明 Table 对象时，将使用模式和表名称的组合把它存储在内部 MetaData 命名空间中。通过搜索键'remote_banks.financial_info'可以在 MetaData.tables 集合中查看此内容：

```
>>> metadata.tables['remote_banks.financial_info']
Table('financial_info', MetaData(),
Column('id', Integer(), table=<financial_info>, primary_key=True, nullable=False),
Column('value', String(length=100), table=<financial_info>, nullable=False),
schema='remote_banks')
```

当引用表与 ForeignKey 或 ForeignKeyConstraint 对象一起使用时，即使引用表也位于同一模式中，也必须使用该带点号的名称：

```
customer = Table(
    "customer",
    metadata,
    Column('id', Integer, primary_key=True),
    Column('financial_info_id', ForeignKey("remote_banks.financial_info.id")),
    schema='remote_banks'
)
```

通过将 MetaData.schema 参数传递到顶级 MetaData 构造器，MetaData 对象还可以为所有 Table.schema 参数设置一个显式的默认选项。使用 MetaData 指定默认模式名称的示例代码如下：

```
metadata = MetaData(schema="remote_banks")

financial_info = Table(
    'financial_info',
    metadata,
    Column('id', Integer, primary_key=True),
    Column('value', String(100), nullable=False),
)
```

5.2.3　插入和查询

插入数据是个有趣的功能。从目标表构建一个 Insert 实例。代码如下：

```
>>> users = Table('users', metadata,
...     Column('id', Integer, primary_key=True),
...     Column('name', String),
...     Column('fullname', String),
...)
>>> ins = users.insert().values(name='jack', fullname='Jack Jones')
>>> str(ins)
'INSERT INTO users (name, fullname) VALUES (:name, :fullname)'
>>> ins.compile().params
```

为了执行插入 SQL 语句，首先需要获得数据库连接。创建的引擎对象是用于数据库连接的存储库。为了获取数据库连接，将使用 engine.connect()方法：

```
>>> conn = engine.connect()
>>> conn
<sqlalchemy.engine.base.Connection object at 0x...>
```

Connection 对象代表一个主动检出的 DBAPI 连接资源。向其提供 Insert 对象，看看会发生什么：

```
>>> result = conn.execute(ins)
INSERT INTO users (name, fullname) VALUES (?, ?)
('jack', 'Jack Jones')
COMMIT
```

创建一个通用的 INSERT 语句，然后以"正常"方式使用它：

```
>>> ins = users.insert()
>>> conn.execute(ins, id=2, name='wendy', fullname='Wendy Williams')
INSERT INTO users (id, name, fullname) VALUES (?, ?, ?)
(2, 'wendy', 'Wendy Williams')
COMMIT
<sqlalchemy.engine.result.ResultProxy object at 0x...>
```

执行多条语句添加一些电子邮件地址的示例如下：

```
>>> conn.execute(addresses.insert(), [
...     {'user_id': 1, 'email_address' : 'jack@yahoo.com'},
...     {'user_id': 1, 'email_address' : 'jack@msn.com'},
...     {'user_id': 2, 'email_address' : 'www@www.org'},
...     {'user_id': 2, 'email_address' : 'wendy@aol.com'},
... ])
INSERT INTO addresses (user_id, email_address) VALUES (?, ?)
((1, 'jack@yahoo.com'), (1, 'jack@msn.com'), (2, 'www@www.org'), (2, 'wendy@aol.com'))
COMMIT
<sqlalchemy.engine.result.ResultProxy object at 0x...>
```

查询这些数据更有趣。通过函数 select() 生成 SELECT 语句。示例代码如下：

```
>>> from sqlalchemy.sql import select
>>> s = select([users])
>>> result = conn.execute(s)
SELECT users.id, users.name, users.fullname
FROM users
()
>>> for row in result:
...     print(row)
(1, u'jack', u'Jack Jones')
(2, u'wendy', u'Wendy Williams')
```

增加查询条件：

```
s = select([users]).where(users.c.name == 'jack')
```

选择特定列的示例代码如下：

```
s = select(users.c.name, users.c.fullname)
```

5.3 Flask-SQLAlchemy 扩展

Flask-SQLAlchemy 是 Flask 的扩展，它为应用程序添加了对 SQLAlchemy 的支持，旨在通过提供有用的默认值和额外的帮助程序来简化 SQLAlchemy 与 Flask 的结合使用，从而更轻松地完成常见任务。

拥有一个 Flask 应用程序的常见情况是：创建 Flask 应用程序，加载选择的配置，然后通过将应用程序传递给 SQLAlchemy 对象来创建这个对象。

创建后，该对象将包含来自 sqlalchemy 和 sqlalchemy.orm 的所有函数和助手。此外，它还提供了一个名为 Model 的类，它是一个声明性基类，可用于声明模型：

```python
from flask import Flask
from flask_sqlalchemy import SQLAlchemy

app = Flask(__name__)
app.config['SQLALCHEMY_DATABASE_URI'] = 'sqlite:////tmp/test.db'
db = SQLAlchemy(app)

class User(db.Model):
    id = db.Column(db.Integer, primary_key=True)
    username = db.Column(db.String(80), unique=True, nullable=False)
    email = db.Column(db.String(120), unique=True, nullable=False)

    def __repr__(self):
        return '<User %r>' % self.username
```

要创建初始数据库，只需从交互式 Python shell 导入 db 对象并运行 SQLAlchemy.create_all() 方法来创建表和数据库：

```
>>> from yourapplication import db
>>> db.create_all()
```

现在创建一些用户：

```
>>> from yourapplication import User
>>> admin = User(username='admin', email='admin@example.com')
>>> guest = User(username='guest', email='guest@example.com')
```

但是它们还没有在数据库中，所以要确保它们是在数据库中：

```
>>> db.session.add(admin)
>>> db.session.add(guest)
>>> db.session.commit()
```

访问数据库中的数据很简单：

```
>>> User.query.all()
[<User u'admin'>, <User u'guest'>]
>>> User.query.filter_by(username='admin').first()
<User u'admin'>
```

第 6 章

Elasticsearch 分布式搜索引擎

Elasticsearch 是一个分布式和高可用的搜索引擎。本章介绍使用 Elasticsearch 实现分布式搜索引擎。

6.1 实现用户界面

首先介绍使用 Node.js 搭建一个简单的用户界面,然后介绍结合 REST API 的搜索界面。

6.1.1 搭建 JavaScript 环境

首先安装 Node.js（https://nodejs.org/），然后创建项目文件夹：

```
>mkdir my-project
>cd my-project
```

使用 Node.js 创建 Web 服务器：

```
var http = require('http');
var fs = require('fs');
var url = require('url');

// Create a server
http.createServer( function (request, response) {
   // Parse the request containing file name
   var pathname = url.parse(request.url).pathname;
   
   // Print the name of the file for which request is made.
   console.log("Request for " + pathname + " received.");
   
   // Read the requested file content from file system
   fs.readFile(pathname.substr(1), function (err, data) {
      if (err) {
```

```
            console.log(err);
            // HTTP Status: 404 : NOT FOUND
            // Content Type: text/plain
            response.writeHead(404, {'Content-Type': 'text/html'});
        }else{
            //Page found
            // HTTP Status: 200 : OK
            // Content Type: text/html
            response.writeHead(200, {'Content-Type': 'text/html;charset=utf-8'});

            // Write the content of the file to response body
            response.write(data.toString());
        }
        // Send the response body
        response.end();
    });
}).listen(8081);
// Console will print the message
console.log('Server running at http://127.0.0.1:8081/');
```

搜索首页内容如下:

```
<html>
<head>
  <title>猎兔搜索</title>
  <META http-equiv=Content-Type content="text/html; charset=utf-8">
</head>
<body>
 <form id="sform" method="get" action="search">
              <input  type="text" name="q"/>
              <input  type=submit value=猎兔搜索>
 </form>
</body>
</html>
```

使用 Handlebars 模板引擎生成搜索结果页。首先在项目中安装 handlebars 包。

```
>npm install handlebars
```

模板文件 searchResult.hbs 内容如下:

```
<!DOCTYPE html>
<html>
<head>
    <meta http-equiv="Content-Type" content="text/html; charset=UTF-8">
    <meta name="viewport" content="width=device-width, initial-scale=1">
    <meta http-equiv="X-UA-Compatible" content="IE=edge">
    <title>搜索结果</title>
```

```html
        <link href="https://maxcdn.bootstrapcdn.com/bootstrap/4.3.1/css/bootstrap.
in.css" rel="stylesheet">
        <link href="../css/style.css" rel="stylesheet" type="text/css">
        <link href="../css/font-awesome.min.css" rel="stylesheet" type="text/css">
    </head>
    <body>

{{#each result}}
    <div id="result" class="row">
        <ol class="list-unstyled col-md-8">
            <li>
                        <h3 class="title text-truncate">
                            <a class="link" href="{{this.url}}">{{this.title}}</a>
                        </h3>
                        <div class="body">
                            <div class="description">{{this.body}}</div>
                        </div>
            </li>
        </ol>
    </div>
{{/each}}

    </body>
</html>
```

提供搜索结果测试数据的 searchResult.json 文件内容如下:

```json
{
  "query":"math",
  "totalHits":307,
  "result" : [
      {
        "url": "http://qq.com",
        "title": "QQ",
        "body": "content"
      },
      {
        "url": "http://patfun.com",
        "title": "健趣",
        "body": "content"
      }
  ]
}
```

searchResult.js 内容如下:

```js
var handlebars = require('handlebars');
var fs = require('fs');
```

```javascript
// get your data into a variable
var searchJson = require('./searchResult.json');

// read the file and use the callback to render
fs.readFile('searchResult.hbs', function(err, data){
  if (!err) {
    // make the buffer into a string
    var source = data.toString();
    // call the render function
    var outputString = renderToString(source, searchJson);
    fs.writeFile('searchResult.html', outputString,    function (err) {
  if (err) {
    return console.log(err);
  }
});


  } else {
    // handle file read error
  }
});

// this will be called after the file is read
function renderToString(source, data) {
  var template = handlebars.compile(source);
  var outputString = template(data);
  return outputString;
}
```

执行 searchResult.js 得到搜索结果页：

```
>node searchResult.js
```

提供 REST API 搜索结果的 Python 代码如下：

```python
#encoding: UTF-8
import json
from flask import Flask, jsonify
from flask import request
from elasticsearch import Elasticsearch

es = Elasticsearch()
app = Flask(__name__)
app.config['JSON_AS_ASCII'] = False
@app.route('/')
def index():
    q = request.args.get('q')
    offset = request.args.get('offset')
```

```python
    res= es.search(index='sites',body={'query':{'match':{'body':q}}})
    print("match body Got %d Hits:" % res['hits']['total']['value'])
    print(res['hits']['hits'])

    jsons = res['hits']['hits']
    items = []
    for hits in jsons:
        itemDict = {}
        itemDict["url"] = hits["_source"]["url"]
        print(hits["_source"]["url"])
        itemDict["title"] = hits["_source"]["title"]
        itemDict["body"] = hits["_source"]["body"]
        items.append(itemDict)

    return jsonify(q=q,hits=items);

app.run()
```

请求 REST API 的代码如下：

```javascript
var request = require('sync-request');
function reqResult(query){
    var url = "http://localhost:5000/?"+ new URLSearchParams({
        q: query
    });
    var res = request('GET', url);
    var ret = res.getBody().toString();
    return ret;
}
```

实现搜索的完整代码如下：

```javascript
var http = require('http');
var fs = require('fs');
const { URL } = require('url');
var request = require('sync-request');
const fetch = require('node-fetch');

var handlebars = require('handlebars');
var fs = require('fs');

// Create a server
http.createServer( function (req, response) {
    var addr = req.url;
    var baseURL = 'http://' + req.headers.host + '/';
    var q = new URL(addr, baseURL);
    // Parse the request containing file name
```

```javascript
    var pathname = q.pathname;

    // Print the name of the file for which request is made.
    console.log("Request for " + pathname + " received.");

    if(pathname.startsWith('/search')){
    var localStr = reqResult(q.searchParams.get('q'));
    var searchJson = JSON.parse(localStr);

    // read the file and use the callback to render
    fs.readFile('searchResult.hbs', function(err, data){
      if (!err) {
        // make the buffer into a string
        var source = data.toString();
        // call the render function
        var outputString = renderToString(source, searchJson);
        response.writeHead(200, {'Content-Type': 'text/html;charset=UTF-8'});
        response.write(outputString);
      } else {
        // handle file read error
      }
    });
    }else{
    // Read the requested file content from file system
    fs.readFile(pathname.substr(1), function (err, data) {
      if (err) {
        console.log(err);
        // HTTP Status: 404 : NOT FOUND
        // Content Type: text/plain
        response.writeHead(404, {'Content-Type': 'text/html'});
      }else{
        //Page found
        // HTTP Status: 200 : OK
        // Content Type: text/plain
        response.writeHead(200, {'Content-Type': 'text/html;charset=UTF-8'});
        // Write the content of the file to response body
        response.write(data.toString());
      }
      // Send the response body
      response.end();
    });
    }
}).listen(8081);
// Console will print the message
console.log('Server running at http://127.0.0.1:8081/');
```

6.1.2 Node.js 基础

Node.js 是能够在服务器端运行 JavaScript 的开放源代码、跨平台 JavaScript 运行时环境。

Node.js 具有 4 种核心数据类型，即数字、布尔值、字符串和对象，还有两种特殊的数据类型，即函数和数组。

这里通过一个自动机的例子介绍 Node.js 面向对象编程。

在 State.js 文件中定义 State 类：

```
class State {
  static maxId = 0
  constructor(accept=false) {
    this.accept = accept
    this.transitions = new Set()
    this.id = State.maxId+1
    State.maxId = State.maxId+1
  }

  addTransition(t){
    this.transitions.add(t)
  }

  step(c){
    console.log("step")
    for(var t of this.transitions){
        console.log("t.min "+t.min + " c: "+c + " t.max "+t.max)
        if (t.min <= c && c <= t.max){
            console.log("enter ...")
            return t.to
        }
    }
    return null
  }

  toString() {
    console.log("toString called")
    var tranStr = "toString "+this.id+" : "

    if (this.accept){
        tranStr += " [accept]"
    }
    else{
        tranStr += " [reject]"
    }

    for(var x of this.transitions){
```

```
            console.log(x)
            tranStr += x+' '
        }
        return tranStr
    }

}

module.exports = State
```

测试 State 类：

```
var state1= new State()

console.log(state1.accept)
console.log(state1.id)

var state2= new State(true)

console.log(state2.accept)
console.log(state2.id)
```

在 Transition.js 文件中引用 State 类。

```
const State = require('./State')

class Transition {
  constructor(mi,mx, to) {
    this.min = mi      //从当前状态转换到下个状态可以接收的字符
    this.max = mx
    this.to = to       //下个状态
  }

  toString() {
    return this.min + " : "+this.max
  }
}

module.exports = Transition
```

这里用到了 require 模块。

表示有限状态机的 Automaton 类实现如下：

```
const State = require('./State');
const Transition = require('./Transition');
const FIFOQueue = require("./FIFOQueue.js")

class Automaton{
  constructor(ini) {
```

```
    this.initial = ini
    this.deterministic = true
  }

  //返回所有的可接受状态
  getAcceptStates(){
    var accepts = new Set()
    var visited = new Set()            //已经遍历过的状态集合
    var worklist = new FIFOQueue()     //工作队列
    worklist.enqueue(this.initial)
    visited.add(this.initial)

    while (worklist.size()>0){
      var s = worklist.dequeue()
      if (s.accept) {
        accepts.add(s)
      }

      for(var t of s.transitions){
        if(!visited.has(t.to)){
          visited.add(t.to)
          worklist.enqueue(t.to)
        }
      }
    }
    return accepts
  }
}

module.exports = Automaton
```

BasicAutomata 类中包括一些方便创建有限状态机的辅助方法：

```
const State = require('./State')
const Transition = require('./Transition')
const Automaton = require('./Automaton')

class BasicAutomata {
  static makeCharRangePlus(min, max) {
    var s1 = new State()
    var s2 = new State(true)

    var minCode = min.charCodeAt(0)
    var maxCode = max.charCodeAt(0)

    var t = new Transition(minCode, maxCode, s2)
```

```
    s1.transitions.add(t)
    s2.transitions.add(t)
    var a = new Automaton(s1)
    return a
  }
}

module.exports = BasicAutomata
```

执行有限状态机的 BasicOperations 类实现如下：

```
const State = require('./State')
const Transition = require('./Transition')
const Automaton = require('./Automaton')

class BasicOperations {
  static run(a, s) {
    var p = a.initial
    for(var c of s){
        var code = c.charCodeAt(0)
        var q = p.step(code)

        if (q == null){
            return false
        }
        p = q
    }
    return p.accept
  }
}

module.exports = BasicOperations
```

测试 BasicOperations：

```
const State = require('./State')
const Transition = require('./Transition')
const Automaton = require('./Automaton')
const BasicAutomata = require('./BasicAutomata')
const BasicOperations = require('./BasicOperations')

//创建有限状态机
a = BasicAutomata.makeCharRangePlus('0','9')

text = '2020'
//检查有限状态机是否能接受文本
var ret = BasicOperations.run(a,text)
console.log(ret)
```

接下来，可以将项目初始化为 npm 项目：

```
>npm init -y
```

将项目初始化为 npm 项目，这为项目带来了两个好处：一是可以将 npm 的库安装到项目中；二是可以添加 npm 脚本，以便在项目生命周期的后期启动、测试或部署项目。首先安装 npm 随附的 Node.js，然后才能在命令行上使用 npm。之后，可以在命令行上初始化 npm 项目：

```
npm config list

npm set init.author.name "<Your Name>"
npm set init.author.email "you@example.com"
npm set init.author.url "example.com"
npm set init.license "MIT"
```

设置 npm 项目后，可以使用 npm 将库安装到项目中。接下来，在命令行上或在编辑器/IDE/浏览器中，为项目的源代码创建 src/文件夹。在此文件夹中，创建 src/index.js 文件作为项目的入口点。然后，在 src/index.js 文件中引入一条日志记录语句，以确保设置可行：

```
console.log('Hello Project.');
```

可以从项目的根文件夹中使用 Node.js 运行此文件：

```
>node src/index.js
```

执行脚本后，日志记录应出现在命令行中。接下来，将此脚本移到 package.json 文件中。package.json 文件内容如下：

```
{
  ...
  "scripts": {
    "start": "node src/index.js",
    "test": "echo \"Error: no test specified\" && exit 1"
  },
  ...
}
```

在命令行上，再次运行脚本，但是这次以如下命令启动。

```
>npm start
```

6.1.3　使用 React 前端库

React 是 Facebook 公司开发的前端库。它用于处理 Web 和移动应用程序的视图层。ReactJS 允许创建可重用的 UI 组件。

首先介绍 React 开发环境搭建。ReactJS 代码可以在 NodeJS 环境中执行。安装完 NodeJS（https://nodejs.org/）后，使用如下命令验证 NPM（Node Package Manager）。NPM 是 Node.js

自带的包管理器。

```
>npm --version
6.12.0
```

接着使用 npm 命令安装 create-react-app 包：

```
>npm install -g create-react-app
```

只需运行以下命令即可创建新应用：

```
>create-react-app my-react-tutorial-app
```

生成出来的 package.json 文件内容如下：

```
{
  "name": "my-react-tutorial-app",
  "version": "0.1.0",
  "private": true,
  "dependencies": {
    "@testing-library/jest-dom": "^5.11.9",
    "@testing-library/react": "^11.2.5",
    "@testing-library/user-event": "^12.7.0",
    "react": "^17.0.1",
    "react-dom": "^17.0.1",
    "react-scripts": "4.0.2",
    "web-vitals": "^1.1.0"
  },
  "scripts": {
    "start": "react-scripts start",
    "build": "react-scripts build",
    "test": "react-scripts test",
    "eject": "react-scripts eject"
  },
  "eslintConfig": {
    "extends": [
      "react-app",
      "react-app/jest"
    ]
  },
  "browserslist": {
    "production": [
      ">0.2%",
      "not dead",
      "not op_mini all"
    ],
    "development": [
      "last 1 chrome version",
      "last 1 firefox version",
```

```
      "last 1 safari version"
    ]
  }
}
```

package.json 文件具有以下属性：

- name：包含传递给 create-react-app 的应用名称。
- version：显示当前版本。
- dependencies：显示应用程序所需的所有必需模块/版本。默认情况下，npm 将安装最新的主版本。
- devDependencies：显示用于在开发环境中测试应用程序的所有模块/版本。
- scripts：具有可用于高效访问 react-scripts 命令的别名。例如，在命令行中运行 npm build，实际上将在后台运行"react-scripts build"。

使用以下命令运行应用程序：

```
>npm start
```

这将在开发模式下运行应用程序。可以在任何浏览器中导航到 http://localhost:3000 来实时预览应用程序。只要在源文件中检测到任何代码更改，该页面就会自动重新加载。在控制台中也可以看到警告和错误。

运行以下命令来构建应用程序：

```
>npm run build
```

开发人员经常将 JavaScript 和 CSS 分解为单独的文件。单独的文件让开发人员可以专注于编写更多只完成一件事情的模块化的代码块。只做一件事的文件会减少开发人员的认知负担，因为维护它们是一项非常烦琐的任务。

当需要将应用程序移至生产环境时，拥有多个 JavaScript 或 CSS 文件并不理想。当用户访问网站时，网站的每个文件都需要一个额外的 HTTP 请求，从而使网站加载速度变慢。因此，为了解决这个问题，可以创建应用程序的"构建"，这样会将所有的 CSS 文件合并到一个文件中，并对 JavaScript 执行相同的操作。这样就可以最大限度地减少用户获取的文件数量和大小。要创建此"构建"，需要使用"构建工具"。因此使用 npm run build。

只有高流量站点（或很快会成为高流量的站点）才需要生成项目的构建版本。如果只是在学习开发，或者只制作流量非常低的网站，那么生成构建可能不值得花时间。

如下命令将以交互方式运行测试。

```
>npm test
```

使用 create-react-app 创建新应用时，所有构建设置实际上都是由该工具预先配置的。因此，无法对构建设置进行任何更新。例如，无法访问 webpack.config 文件。它实际上是由"react-scripts"构建依赖项管理的。但是，有一种方法可以自定义设置，而不受 Create React App 提供的配置的限制。

弹出能够完全控制配置文件以及 webpack / Babel / ESLint 之类的依赖项。实际弹出会创建 Create React 应用程序配置，并将其移动到应用程序中。运行弹出命令后，可以看到在项目中创建的"config"文件夹，其中包含用于开发和生产的 webpack.config 之类的文件以及一个 webpackDevServer.config 文件。另外，可以看到在 package.json 中，单个构建依赖关系 react-scripts 已从教程项目中删除，并且列出了所有单独的依赖关系。

请注意，运行弹出是不可逆的步骤或动作。也就是说，执行此命令后，将无法还原或返回到初始状态。

弹出后，可以继续正常使用应用程序，并且上面讨论的所有命令都将起作用（构建、启动和测试）。但是现在要负责配置。因此，如果对配置或依赖项了解不多，最好避免使用弹出。

对于弹出，只需执行以下命令：

```
>npm run eject
```

接下来将构建一个 React 应用程序：
- 有"登录/注销"和"注册"页面。
- 表单数据在发送到后端之前，将由前端进行验证。
- 导航栏会根据用户的角色（管理员、主持人、用户）自动更改其项目。

如下命令建立 React.js 项目：

```
>npx create-react-app react-jwt-auth
```

运行如下命令来添加 React 路由器：

```
>npm install react-router-dom
```

打开 src/index.js 并通过 BrowserRouter 对象包装 App 组件。

```
import React from "react";
import ReactDOM from "react-dom";
import { BrowserRouter } from "react-router-dom";

import App from "./App";
import * as serviceWorker from "./serviceWorker";

ReactDOM.render(
  <BrowserRouter>
    <App />
  </BrowserRouter>,
  document.getElementById("root")
);

serviceWorker.unregister();
```

运行如下命令来导入 Bootstrap：

```
> npm install bootstrap
```

打开 src/App.js 并按照以下内容修改其中的代码：

```
import React, { Component } from "react";
import "bootstrap/dist/css/bootstrap.min.css";

class App extends Component {
  render() {
    // ...
  }
}


export default App;
```

在 src/services 文件夹中创建身份验证服务和数据服务。开发这些服务之前，需要使用以下命令安装 Axios：

```
>npm install axios
```

认证服务将 Axios 用于 HTTP 请求，将本地存储用于用户信息和 JWT。它提供以下重要方法：

- login()：POST {username, password}并且保存 JWT 到本地存储。
- logout()：从本地存储中删除 JWT。
- register()：POST {username, email, password}。
- getCurrentUser()：获取存储的用户信息（包括 JWT）。

认证服务实现如下：

```
import axios from "axios";

const API_URL = "http://localhost:8080/api/auth/";

class AuthService {
  login(username, password) {
    return axios
      .post(API_URL + "signin", {
        username,
        password
      })
      .then(response => {
        if (response.data.accessToken) {
          localStorage.setItem("user", JSON.stringify(response.data));
        }

        return response.data;
```

```
    });
  }

  logout() {
    localStorage.removeItem("user");
  }

  register(username, email, password) {
    return axios.post(API_URL + "signup", {
      username,
      email,
      password
    });
  }

  getCurrentUser() {
    return JSON.parse(localStorage.getItem('user'));;
  }
}

export default new AuthService();
```

还提供了从服务器检索数据的方法。在访问受保护资源的情况下，HTTP 请求需要 Authorization 标头。

在 auth-header.js 中创建一个名为 authHeader() 的辅助函数：

```
export default function authHeader() {
  const user = JSON.parse(localStorage.getItem('user'));

  if (user && user.accessToken) {
    return { Authorization: 'Bearer ' + user.accessToken };
  } else {
    return {};
  }
}
```

上面的代码检查本地存储中的用户项。如果存在具有 accessToken（JWT）的登录用户，则返回 HTTP Authorization 标头；否则，返回一个空对象。

现在，在 user.service.js 中定义用于访问数据的服务：

```
import axios from 'axios';
import authHeader from './auth-header';

const API_URL = 'http://localhost:8080/api/test/';

class UserService {
  getPublicContent() {
```

```
    return axios.get(API_URL + 'all');
  }

  getUserBoard() {
    return axios.get(API_URL + 'user', { headers: authHeader() });
  }

  getModeratorBoard() {
    return axios.get(API_URL + 'mod', { headers: authHeader() });
  }

  getAdminBoard() {
    return axios.get(API_URL + 'admin', { headers: authHeader() });
  }
}

export default new UserService();
```

可以看到,当请求授权资源时,在函数 authHeader() 的帮助下添加了 HTTP 标头。

在 src 文件夹中,创建一个名为 components 的新文件夹,并添加 3 个文件:login.component.js、register.component.js 和 profile.component.js。

现在需要一个用于表单验证的库,因此要向项目中添加 react-validation 库。运行命令:

```
>npm install react-validation validator
```

如果要在此示例中使用 react-validation,则需要导入以下项目:

```
import Form from "react-validation/build/form";
import Input from "react-validation/build/input";
import CheckButton from "react-validation/build/button";

import { isEmail } from "validator";
```

还使用验证程序中的函数 isEmail() 来验证电子邮件。

这就是将它们放入带有 validations 属性的 render() 方法中的方式:

```
const required = value => {
  if (!value) {
    return (
      <div className="alert alert-danger" role="alert">
        This field is required!
      </div>
    );
  }
};

const email = value => {
  if (!isEmail(value)) {
```

```
      return (
        <div className="alert alert-danger" role="alert">
          This is not a valid email.
        </div>
      );
    }
  };

  render() {
    return (
    ...
      <Form
        onSubmit={this.handleLogin}
        ref={c => {this.form = c;}}
      >
        ...
        <Input
          type="text"
          className="form-control"
          ...
          validations={[required, email]}
        />

        <CheckButton
          style={{ display: "none" }}
          ref={c => {this.checkBtn = c;}}
        />
      </Form>
    ...
    );
  }
```

将调用表单的 validateAll() 方法来检查验证功能。然后 CheckButton 帮助验证表单验证是否成功。因此，此按钮将不会显示在表单上。

```
this.form.validateAll();

if (this.checkBtn.context._errors.length === 0) {
  // do something when no error
}
```

登录页面有一个带有用户名和密码的表单。

将它们当作必填字段来验证。

如果验证通过，将调用 AuthService.login() 方法，然后将用户定向到"个人资料"页面："this.props.history.push("/profile");"，或显示带有响应错误的消息。

login.component.js 内容如下：

```
import React, { Component } from "react";
import Form from "react-validation/build/form";
import Input from "react-validation/build/input";
import CheckButton from "react-validation/build/button";

import AuthService from "../services/auth.service";

const required = value => {
  if (!value) {
    return (
      <div className="alert alert-danger" role="alert">
        This field is required!
      </div>
    );
  }
};

export default class Login extends Component {
  constructor(props) {
    super(props);
    this.handleLogin = this.handleLogin.bind(this);
    this.onChangeUsername = this.onChangeUsername.bind(this);
    this.onChangePassword = this.onChangePassword.bind(this);

    this.state = {
      username: "",
      password: "",
      loading: false,
      message: ""
    };
  }

  onChangeUsername(e) {
    this.setState({
      username: e.target.value
    });
  }

  onChangePassword(e) {
    this.setState({
      password: e.target.value
    });
  }

  handleLogin(e) {
    e.preventDefault();
```

```
    this.setState({
      message: "",
      loading: true
    });

    this.form.validateAll();

    if (this.checkBtn.context._errors.length === 0) {
      AuthService.login(this.state.username, this.state.password).then(
        () => {
          this.props.history.push("/profile");
          window.location.reload();
        },
        error => {
          const resMessage =
            (error.response &&
              error.response.data &&
              error.response.data.message) ||
            error.message ||
            error.toString();

          this.setState({
            loading: false,
            message: resMessage
          });
        }
      );
    } else {
      this.setState({
        loading: false
      });
    }
  }

  render() {
    return (
      <div className="col-md-12">
        <div className="card card-container">
          <img
            src="//ssl.gstatic.com/accounts/ui/avatar_2x.png"
            alt="profile-img"
            className="profile-img-card"
          />

          <Form
```

```jsx
    onSubmit={this.handleLogin}
    ref={c => {
      this.form = c;
    }}
  >
    <div className="form-group">
      <label htmlFor="username">Username</label>
      <Input
        type="text"
        className="form-control"
        name="username"
        value={this.state.username}
        onChange={this.onChangeUsername}
        validations={[required]}
      />
    </div>

    <div className="form-group">
      <label htmlFor="password">Password</label>
      <Input
        type="password"
        className="form-control"
        name="password"
        value={this.state.password}
        onChange={this.onChangePassword}
        validations={[required]}
      />
    </div>

    <div className="form-group">
      <button
        className="btn btn-primary btn-block"
        disabled={this.state.loading}
      >
        {this.state.loading && (
          <span className="spinner-border spinner-border-sm"></span>
        )}
        <span>Login</span>
      </button>
    </div>

    {this.state.message && (
      <div className="form-group">
        <div className="alert alert-danger" role="alert">
          {this.state.message}
        </div>
```

```
        </div>
      )}
      <CheckButton
        style={{ display: "none" }}
        ref={c => {
          this.checkBtn = c;
        }}
      />
    </Form>
  </div>
</div>
    );
  }
}
```

注册页面类似于登录页面。对于表单验证,有更多详细信息:

- username:必填,3~20 个字符。
- email:必填,电子邮件格式。
- password:必填,6~40 个字符。

调用 AuthService.register()方法并显示响应消息(成功或错误)。

register.component.js 文件内容如下:

```
import React, { Component } from "react";
import Form from "react-validation/build/form";
import Input from "react-validation/build/input";
import CheckButton from "react-validation/build/button";
import { isEmail } from "validator";

import AuthService from "../services/auth.service";

const required = value => {
  if (!value) {
    return (
      <div className="alert alert-danger" role="alert">
        This field is required!
      </div>
    );
  }
};

const email = value => {
  if (!isEmail(value)) {
    return (
      <div className="alert alert-danger" role="alert">
        This is not a valid email.
```

```jsx
      </div>
    );
  }
};

const vusername = value => {
  if (value.length < 3 || value.length > 20) {
    return (
      <div className="alert alert-danger" role="alert">
        The username must be between 3 and 20 characters.
      </div>
    );
  }
};

const vpassword = value => {
  if (value.length < 6 || value.length > 40) {
    return (
      <div className="alert alert-danger" role="alert">
        The password must be between 6 and 40 characters.
      </div>
    );
  }
};

export default class Register extends Component {
  constructor(props) {
    super(props);
    this.handleRegister = this.handleRegister.bind(this);
    this.onChangeUsername = this.onChangeUsername.bind(this);
    this.onChangeEmail = this.onChangeEmail.bind(this);
    this.onChangePassword = this.onChangePassword.bind(this);

    this.state = {
      username: "",
      email: "",
      password: "",
      successful: false,
      message: ""
    };
  }

  onChangeUsername(e) {
    this.setState({
      username: e.target.value
    });
```

```javascript
  }

  onChangeEmail(e) {
    this.setState({
      email: e.target.value
    });
  }

  onChangePassword(e) {
    this.setState({
      password: e.target.value
    });
  }

  handleRegister(e) {
    e.preventDefault();

    this.setState({
      message: "",
      successful: false
    });

    this.form.validateAll();

    if (this.checkBtn.context._errors.length === 0) {
      AuthService.register(
        this.state.username,
        this.state.email,
        this.state.password
      ).then(
        response => {
          this.setState({
            message: response.data.message,
            successful: true
          });
        },
        error => {
          const resMessage =
            (error.response &&
              error.response.data &&
              error.response.data.message) ||
            error.message ||
            error.toString();

          this.setState({
            successful: false,
```

```jsx
            message: resMessage
          });
        }
      );
    }
  }

  render() {
    return (
      <div className="col-md-12">
        <div className="card card-container">
          <img
            src="//ssl.gstatic.com/accounts/ui/avatar_2x.png"
            alt="profile-img"
            className="profile-img-card"
          />

          <Form
            onSubmit={this.handleRegister}
            ref={c => {
              this.form = c;
            }}
          >
            {!this.state.successful && (
              <div>
                <div className="form-group">
                  <label htmlFor="username">Username</label>
                  <Input
                    type="text"
                    className="form-control"
                    name="username"
                    value={this.state.username}
                    onChange={this.onChangeUsername}
                    validations={[required, vusername]}
                  />
                </div>

                <div className="form-group">
                  <label htmlFor="email">Email</label>
                  <Input
                    type="text"
                    className="form-control"
                    name="email"
                    value={this.state.email}
                    onChange={this.onChangeEmail}
                    validations={[required, email]}
```

```jsx
              />
            </div>

            <div className="form-group">
              <label htmlFor="password">Password</label>
              <Input
                type="password"
                className="form-control"
                name="password"
                value={this.state.password}
                onChange={this.onChangePassword}
                validations={[required, vpassword]}
              />
            </div>

            <div className="form-group">
              <button className="btn btn-primary btn-block">Sign Up</button>
            </div>
          </div>
        )}

        {this.state.message && (
          <div className="form-group">
            <div
              className={
                this.state.successful
                  ? "alert alert-success"
                  : "alert alert-danger"
              }
              role="alert"
            >
              {this.state.message}
            </div>
          </div>
        )}
        <CheckButton
          style={{ display: "none" }}
          ref={c => {
            this.checkBtn = c;
          }}
        />
      </Form>
    </div>
  </div>
 );
}
```

}
```

个人资料页通过调用 AuthService.getCurrentUser()方法从本地存储获取当前用户，并显示用户信息（带有令牌）。

profile.component.js 内容如下：

```js
import React, { Component } from "react";
import AuthService from "../services/auth.service";

export default class Profile extends Component {
  constructor(props) {
    super(props);

    this.state = {
      currentUser: AuthService.getCurrentUser()
    };
  }

  render() {
    const { currentUser } = this.state;

    return (
      <div className="container">
        <header className="jumbotron">
          <h3>
            <strong>{currentUser.username}</strong> Profile
          </h3>
        </header>
        <p>
          <strong>Token:</strong>{" "}
          {currentUser.accessToken.substring(0, 20)} ...{" "}
          {currentUser.accessToken.substr(currentUser.accessToken.length - 20)}
        </p>
        <p>
          <strong>Id:</strong>{" "}
          {currentUser.id}
        </p>
        <p>
          <strong>Email:</strong>{" "}
          {currentUser.email}
        </p>
        <strong>Authorities:</strong>
        <ul>
          {currentUser.roles &&
            currentUser.roles.map((role, index) => <li key={index}>{role}</li>)}
        </ul>
```
```

```
      </div>
    );
  }
}
```

创建 React 组件以访问资源。这些组件将使用 UserService 从 API 请求数据。

主页是显示公共内容的公共页面。用户无须登录即可查看此页面。

home.component.js 文件内容如下：

```
import React, { Component } from "react";

import UserService from "../services/user.service";

export default class Home extends Component {
  constructor(props) {
    super(props);

    this.state = {
      content: ""
    };
  }

  componentDidMount() {
    UserService.getPublicContent().then(
      response => {
        this.setState({
          content: response.data
        });
      },
      error => {
        this.setState({
          content:
            (error.response && error.response.data) ||
            error.message ||
            error.toString()
        });
      }
    );
  }

  render() {
    return (
      <div className="container">
        <header className="jumbotron">
          <h3>{this.state.content}</h3>
        </header>
```

```
      </div>
    );
  }
}
```

将有 3 个基于角色的页面用于访问受保护的数据:
- BoardUser 页面调用 UserService.getUserBoard()。
- BoardModerator 页面调用 UserService.getModeratorBoard()。
- BoardAdmin 页面调用 UserService.getAdminBoard()。

这里只展示用户页面,其他页面与此页面相似。

board-user.component.js 文件内容如下:

```
import React, { Component } from "react";

import UserService from "../services/user.service";

export default class BoardUser extends Component {
  constructor(props) {
    super(props);

    this.state = {
      content: ""
    };
  }

  componentDidMount() {
    UserService.getUserBoard().then(
      response => {
        this.setState({
          content: response.data
        });
      },
      error => {
        this.setState({
          content:
            (error.response &&
              error.response.data &&
              error.response.data.message) ||
            error.message ||
            error.toString()
        });
      }
    );
  }
```

```
  render() {
    return (
      <div className="container">
        <header className="jumbotron">
          <h3>{this.state.content}</h3>
        </header>
      </div>
    );
  }
}
```

添加导航栏并定义路由。

现在在 App 组件中添加一个导航栏。这是应用程序的根容器。导航栏会根据登录状态和当前用户的角色动态变化。导航栏包括:

- 主页: 总是存在。
- 登录和注册: 如果用户尚未登录。
- 用户: AuthService.getCurrentUser()返回一个值。
- 主持人面板: 包括 ROLE_MODERATOR 的角色。
- 管理员面板: 包括 ROLE_ADMIN 的角色。

src/App.js 文件内容如下:

```
import React, { Component } from "react";
import { BrowserRouter as Router, Switch, Route, Link } from "react-router-dom";
import "bootstrap/dist/css/bootstrap.min.css";
import "./App.css";

import AuthService from "./services/auth.service";

import Login from "./components/login.component";
import Register from "./components/register.component";
import Home from "./components/home.component";
import Profile from "./components/profile.component";
import BoardUser from "./components/board-user.component";
import BoardModerator from "./components/board-moderator.component";
import BoardAdmin from "./components/board-admin.component";

class App extends Component {
  constructor(props) {
    super(props);
    this.logOut = this.logOut.bind(this);

    this.state = {
      showModeratorBoard: false,
      showAdminBoard: false,
```

```
      currentUser: undefined
    };
  }

  componentDidMount() {
    const user = AuthService.getCurrentUser();

    if (user) {
      this.setState({
        currentUser: user,
        showModeratorBoard: user.roles.includes("ROLE_MODERATOR"),
        showAdminBoard: user.roles.includes("ROLE_ADMIN")
      });
    }
  }

  logOut() {
    AuthService.logout();
  }

  render() {
    const { currentUser, showModeratorBoard, showAdminBoard } = this.state;

    return (
      <Router>
        <div>
          <nav className="navbar navbar-expand navbar-dark bg-dark">
            <Link to={"/"} className="navbar-brand">
              bezKoder
            </Link>
            <div className="navbar-nav mr-auto">
              <li className="nav-item">
                <Link to={"/home"} className="nav-link">
                  Home
                </Link>
              </li>

              {showModeratorBoard && (
                <li className="nav-item">
                  <Link to={"/mod"} className="nav-link">
                    Moderator Board
                  </Link>
                </li>
              )}

              {showAdminBoard && (
```

```jsx
            <li className="nav-item">
              <Link to={"/admin"} className="nav-link">
                Admin Board
              </Link>
            </li>
          )}

          {currentUser && (
            <li className="nav-item">
              <Link to={"/user"} className="nav-link">
                User
              </Link>
            </li>
          )}
        </div>

        {currentUser ? (
          <div className="navbar-nav ml-auto">
            <li className="nav-item">
              <Link to={"/profile"} className="nav-link">
                {currentUser.username}
              </Link>
            </li>
            <li className="nav-item">
              <a href="/login" className="nav-link" onClick={this.logOut}>
                LogOut
              </a>
            </li>
          </div>
        ) : (
          <div className="navbar-nav ml-auto">
            <li className="nav-item">
              <Link to={"/login"} className="nav-link">
                Login
              </Link>
            </li>

            <li className="nav-item">
              <Link to={"/register"} className="nav-link">
                Sign Up
              </Link>
            </li>
          </div>
        )}
      </nav>
```

```
      <div className="container mt-3">
        <Switch>
          <Route exact path={["/", "/home"]} component={Home} />
          <Route exact path="/login" component={Login} />
          <Route exact path="/register" component={Register} />
          <Route exact path="/profile" component={Profile} />
          <Route path="/user" component={BoardUser} />
          <Route path="/mod" component={BoardModerator} />
          <Route path="/admin" component={BoardAdmin} />
        </Switch>
      </div>
    </div>
  </Router>
  );
 }
}

export default App;
```

接下来为 React 组件添加 CSS 样式。打开 src/App.css 文件并编写一些 CSS 代码，如下所示：

```
label {
  display: block;
  margin-top: 10px;
}

.card-container.card {
  max-width: 350px !important;
  padding: 40px 40px;
}

.card {
  background-color: #f7f7f7;
  padding: 20px 25px 30px;
  margin: 0 auto 25px;
  margin-top: 50px;
  -moz-border-radius: 2px;
  -webkit-border-radius: 2px;
  border-radius: 2px;
  -moz-box-shadow: 0px 2px 2px rgba(0, 0, 0, 0.3);
  -webkit-box-shadow: 0px 2px 2px rgba(0, 0, 0, 0.3);
  box-shadow: 0px 2px 2px rgba(0, 0, 0, 0.3);
}

.profile-img-card {
  width: 96px;
```

```
    height: 96px;
    margin: 0 auto 10px;
    display: block;
    -moz-border-radius: 50%;
    -webkit-border-radius: 50%;
    border-radius: 50%;
}
```

由于大多数 HTTP Server 使用 CORS（跨域资源共享）配置来接受受限于某些站点或端口的资源共享，因此还需要为应用配置端口。

在项目文件夹中，创建具有以下内容的.env 文件：

```
PORT=8081
```

现在将应用程序设置为在端口 8081 上运行。

6.1.4 使用 webpack 模块捆绑器

webpack 是一个模块捆绑器。它的主要目的是捆绑 JavaScript 文件以供在浏览器中使用，但它也能够转换、捆绑或打包几乎任何资源或资产。

webpack 的入口点是收集前端项目的所有依赖项的起点。实际上，这是一个简单的 JavaScript 文件。webpack 的输出是在构建过程中收集到的生成的 JavaScript 和静态文件的位置。webpack 的默认输出文件夹是 dist。生成的 JavaScript 文件是所谓的捆绑软件的一部分。

加载程序是第三方扩展程序，可帮助 webpack 处理各种文件扩展名。例如，有用于 CSS、图像或 txt 文件的加载器。加载程序的目标是在模块中转换文件（JavaScript 以外的文件）。文件成为模块后，webpack 可以将其用作项目中的依赖项。

插件是第三方扩展，可以更改 webpack 的工作方式。例如，有用于解压 HTML、CSS 或设置环境变量的插件。

webpack 有开发和生产两种操作模式。它们之间的主要区别是生产模式会自动将最小化和其他优化应用于 JavaScript 代码。

代码拆分或延迟加载是一种避免较大捆绑包的优化技术。通过代码拆分，开发人员可以决定仅响应某些用户交互（例如单击或路由更改，或其他条件）来加载整个 JavaScript 代码块。拆分出来的一段代码变成了一个组块。

本地安装 webpack：

```
>npm install --save-dev webpack
>npm install --save-dev webpack-cli
```

这会创建一个新的 package.json 文件来跟踪已安装的软件包。在当前目录下应该会看到一个名为 node_modules 的新文件夹。package.json 文件内容如下：

```json
{
  "name": "my-project",
  "version": "1.0.0",
  "description": "",
  "main": "index.js",
  "scripts": {
    "start": "webpack serve --config ./webpack.config.js --mode development",
    "test": "echo \"Error: no test specified\" && exit 1"
  },
  "keywords": [],
  "author": "",
  "license": "ISC",
  "devDependencies": {
    "webpack": "^5.21.2",
    "webpack-cli": "^4.5.0",
    "webpack-dev-server": "^3.11.2"
  }
}
```

现在已安装 webpack，可以使用以下命令运行它：

```
>npx webpack
```

假设项目的入口点是./src/index，并且希望捆绑文件位于 dist/main.js 中。如果想以不同的方式配置设置或配置加载器，就需要制作一个配置文件。大多数情况下，配置文件将作为 webpack.config.js 放在项目的根目录下。需要使用--config 标志加载它，但可以通过在 package.json 中指定一个脚本来自动执行此操作：

```
"scripts": {
  "build": "webpack --config webpack.config.js"
}
```

现在，无论何时运行：

```
>npm run build
```

webpack 都将运行并捆绑该项目。

6.2　自动完成

为了实现搜索框自动完成功能，搜索首页 static/index.html 中的相关内容如下：

```
<form name="rootAction_index_Form" id="searchForm" method="get" action="./search/">
    <input type="text" name="query" maxlength="1000" size="50" value="" id="contentQuery" autocomplete="off">
    <button type="submit" name="search" id="searchButton">
```

```
        Search
    </button>
</form>
```

static/js/autocomplete.js 文件中的 JavaScript 代码如下:

```
/*
    function: addEvent
    @param: obj        (Object)(Required)
        - 想要附加事件的对象
    @param: type       (String)(Required)
        - 希望建立的事件类型
    @param: callback   (Function)(Required)
        - 希望事件监听器调用的方法
    @param: eventReturn (Boolean)(Optional)
        - 是否要将事件对象返回给回调方法
*/
var addEvent = function(obj, type, callback, eventReturn) {
    if (obj == null || typeof obj === 'undefined')
        return;

    if (obj.addEventListener)
        obj.addEventListener(type, callback, eventReturn ? true : false);
    else
        obj["on" + type] = callback;
};
const keys = {
    ENTER: 13,
    ARROW_UP: 38,
    ARROW_DOWN: 40,
};
var suggestor = new function() {
    var listSelNum = 0;
    var listNum = 0;
    var inputText = "";
    //DOM 树增加 suggestorBox
    var boxElement = document.createElement('div');
    boxElement.setAttribute("id", "suggestorBox");
    boxElement.className = 'suggestorBox';
    boxElement.style.display = "none";         //默认不显示此元素
    boxElement.style.position = "fixed";
    boxElement.style.zIndex = "2";             //设置堆叠顺序
    boxElement.style.setProperty('background-color', 'rgb(255, 255, 255)');
    boxElement.style.width = '278px';
    document.body.appendChild(boxElement);
    //增加 suggestorBox 的显示风格
    var css = '#suggestorBox li:hover { background-color:rgba(82, 168, 236,
```

```javascript
0.1); }';
      var style = document.createElement('style');
      if (style.styleSheet) {
         style.styleSheet.cssText = css;
      } else {
         style.appendChild(document.createTextNode(css));
      }
      document.getElementsByTagName('head')[0].appendChild(style);
      var olEle = document.createElement('ol');
      olEle.style.cssText = 'list-style: none; padding: 0px; margin: 2px;'
      boxElement.appendChild(olEle);
      this.selectListUp = function selectListUp() {
         listSelNum--;
         if (listSelNum < 0) {
            listSelNum = listNum;
         }
      }
      this.selectListDown = function selectListDown() {
         listSelNum++;
         if (listSelNum > listNum) {
            listSelNum = 0;
         }
      }
      var textArea = document.getElementById('contentQuery');
      var watchInput = function(evt) {
         suggestor.inputKeyDown(evt);
      };
      addEvent(textArea, "keydown", watchInput, true);
      var searchForm = document.searchForm;
      this.fixList = function() {
         if (listSelNum > 0) {
            textArea.value =
                     boxElement.children[0].children[listSelNum - 1].innerHTML;
         }
         else {
            textArea.value = inputText;
         }
      }
      this.render = function() {
         for (k = 1; k <= listNum; k++) {
            if (k === listSelNum) {
               boxElement.children[0].children[k - 1].style
                        .setProperty('background-color', 'rgba(82, 168, 236,
0.1)');
            } else {
               boxElement.children[0].children[k - 1].style
```

```javascript
                    .setProperty('background-color', 'rgb(255, 255, 255)');
        }
    }
}
var url = '/autoComplete?q=';
this.wordsList = function () {
    boxElement.style.display = "none";
    //请求后端提供的自动完成接口
    fetch(url + textArea.value)
        .then(res => res.json())
        .then(res => {
            suggestor.createAutoCompleteList(res);
            suggestor.listNum = res.length;
        });
    boxElement.style.display = "block";
    listSelNum = 0;
}
this.inputKeyDown = function inputKeyDown(e) {
    if (e.keyCode === keys.ENTER) {
        document.getElementById('searchButton').click();
    }
    else if (e.keyCode === keys.ARROW_UP) {
        if (boxElement.style.display === "none") {
            return;
        }
        suggestor.selectListUp();
        suggestor.render();
        suggestor.fixList();
    } else if (e.keyCode === keys.ARROW_DOWN) {
        if (boxElement.style.display === "none") {
            return;
        }
        suggestor.selectListDown();
        suggestor.render();
        suggestor.fixList();
    }
    else if (
        (e.keyCode >= 48 && e.keyCode <= 90) ||
        (e.keyCode >= 96 && e.keyCode <= 105) ||
        (e.keyCode >= 186) ||
        e.keyCode === 8 ||
        e.keyCode === 32 ||
        e.keyCode === 46) {
        //使用在setTimeout函数内注册的函数来得到输入框中的实际值
        setTimeout(suggestor.wordsList, 1);
    }
```

```javascript
}
var watchInputPress = function(evt) {
   suggestor.inputKeyPress(evt);
};

addEvent(textArea, "keyup", watchInputPress, true);

this.inputKeyPress = function inputKeyPress(e) {
   if (
      (e.keyCode >= 48 && e.keyCode <= 90) ||
      (e.keyCode >= 96 && e.keyCode <= 105) ||
      (e.keyCode >= 186) ||
      e.keyCode === 8 ||
      e.keyCode === 32 ||
      e.keyCode === 46) {
      inputText = textArea.value;
   }
}
this.createAutoCompleteList = function(reslist) {
   olEle.innerHTML = '';
   for (k = 1; k <= reslist.length; k++) {
      var liEle = document.createElement('li');
      liEle.style.cssText = 'padding: 2px;';
      liEle.setAttribute("id", k);
      liEle.tabIndex = 0;
      liEle.innerHTML = reslist[k - 1];
      liEle.addEventListener("click", function(e) {
          textArea.value = this.innerHTML;
          searchForm.submit();
      });
      liEle.addEventListener("mouseout", function() {
          this.style.setProperty('background-color', 'rgb(255, 255, 255)');
      });
      liEle.addEventListener('mouseenter', function() {
          suggestor.listSelNum = this.id;
          this.style.setProperty('background-color', 'rgba(82, 168, 236, 0.1)');
      }, true);
      liEle.addEventListener('keydown', function(e, me) {
          if (e.keyCode === keys.ARROW_UP) {
             suggestor.listSelNum = this.id;
             suggestor.selectListUp();
             this.style.setProperty('background-color', 'rgb(255, 255, 255)');
             boxElement.children[0].children[suggestor.listSelNum - 1].style
                .setProperty('background-color', 'rgba(82, 168, 236, 0.1)');
             suggestor.fixList();
             textArea.focus();
```

```javascript
            }
            else if (e.keyCode === keys.ARROW_DOWN) {
                suggestor.listSelNum = this.id;
                suggestor.selectListDown();
                this.style.setProperty('background-color', 'rgb(255, 255, 255)');
                boxElement.children[0].children[suggestor.listSelNum - 1].style
                    .setProperty('background-color', 'rgba(82, 168, 236, 0.1)');
                suggestor.fixList();
                textArea.focus();
            }
        }, true);
        olEle.appendChild(liEle);
    }
    listNum = reslist.length;
}
//匿名函数
this.resize = function () {
    var pos = getPosition(textArea);
    boxElement.style.top = (pos.y + textArea.offsetHeight + 6) + 'px';
    boxElement.style.left = pos.x + 'px';
    boxElement.style.height = 'auto';
}
}
suggestor.resize();
var watchResize = function(evt) {
    suggestor.resize();
};
addEvent(window, 'resize', watchResize, true);
//获取元素确切位置的辅助函数
function getPosition(el) {
    var xPos = 0;
    var yPos = 0;
    while (el) {
        if (el.tagName == "BODY") {
            //获取BODY元素的确切位置
            var xScroll = el.scrollLeft || document.documentElement.scrollLeft;
            var yScroll = el.scrollTop || document.documentElement.scrollTop;
            xPos += (el.offsetLeft - xScroll + el.clientLeft);
            yPos += (el.offsetTop - yScroll + el.clientTop);
        } else {
            //对于所有非BODY元素获取确切位置
            xPos += (el.offsetLeft - el.scrollLeft + el.clientLeft);
            yPos += (el.offsetTop - el.scrollTop + el.clientTop);
        }
        el = el.offsetParent;
    }
```

```
    return {
        x: xPos,
        y: yPos
    };
}
```

在网页中引用 autocomplete.js：

```
<script charset="utf-8" type="text/javascript" src="./js/autocomplete.js"></script>
```

后端的 Node.js 中运行的静态页面位于 public 路径。如果还没有安装 Node.js，则首先从 https://nodejs.org/en/download/ 下载操作系统所需要的安装包，然后安装 Node.js。

检查安装：

```
>node -v
v14.15.4
```

express 是一个基于 Node.js 的 Web 应用程序开发框架。安装 express 模块：

```
>npm install express
```

server.js 文件内容如下：

```
// Load Node modules
var express = require('express');
// Initialise Express
var app = express();
// Render static files
app.use(express.static('public'));
// Port website will run on
app.listen(8080);
```

然后在终端输入：npm init。按 Enter 键接受以下所有选项的默认参数，但请确保入口点是 server.js。

启动 express 服务器：

```
>npm start
```

然后在浏览器中转到 localhost:8080/index.html。刚刚创建的 express 服务器现在应该正在提供网站的静态文件。

如下代码通过 autoComplete 端点提供数据：

```
app.get("/autoComplete", (req, res, next) => {
 res.json(["Tony","Lisa","Michael","Ginger","Food"]);
});
```

通过 req.query 对象得到 url "/autoComplete?q=fla" 的查询参数：

```
app.get("/autoComplete", (req, res, next) => {
 console.log(req.query.q);
});
```

6.3 拼写纠错

为了讨论对搜索引擎查询最有效的拼写检查技术，首先看一下单词拼写检查的概率模型：

$$\text{Spell}(w) = \arg\max_{e \in C} P(e|w) = \arg\max_{e \in C} \frac{P(w|e)P(e)}{P(w)} \tag{6-1}$$

对于任何 e 来讲，出现 w 的概率 $P(w)$ 都是一样的，从而可以在式（6-1）中忽略它，写成：

$$\text{Spell}(w) = \arg\max_{e \in C} P(w|e)P(e) \tag{6-2}$$

式（6-2）有 3 部分，从右到左分别是：

$P(e)$：文章中出现一个正确拼写词 e 的概率。也就是说，在英语文章中，e 出现的概率有多大呢？因为这个概率完全由英语这种语言决定，一般称之为语言模型。例如，英语中出现 the 的概率 $P(\text{'the'})$ 就相对高，而出现 $P(\text{'zxzxzxzyy'})$ 的概率接近 0（假设后者也是一个词）。

$P(w|e)$：在用户想键入 e 的情况下键入了 w 的概率。因为这个是代表用户会以多大的概率把 e 键入成 w，因此把它称为误差模型。

$\arg\max_{e \in C}$：用来枚举所有可能的 e 并且选取概率最大的那个词。因为有理由相信，一个正确的单词出现的频率高，用户又容易把它键入成另一个错误的单词，那么，那个键入错的单词应该被更正为这个正确的。

为什么把最简单的一个 $P(e|w)$ 变成两项复杂的式子来计算？因为 $P(e|w)$ 是和这两项同时相关的，因此拆成两项反而容易处理。例如，一个单词 thew 拼错了，看上去 thaw 应该是正确的，因为就是把 a 键入成 e 了。然而，也有可能用户想要的是 the，因为 the 是英语中常见的一个词，并且很有可能键入时手不小心从 e 滑到 w。因此，在这种情况下，想要计算 $P(e|w)$ 就必须同时考虑 e 出现的概率和从 e 到 w 的概率。把一项拆成两项让这个问题更加容易和清晰。

对于给定词 w 可以通过编辑距离（Edit Distance，也称为莱文斯坦距离）挑选出相似的候选正确词 e 的集合。编辑距离越小，候选正确词越少，计算也越快。76%的正确词和错误词的编辑距离是 1。所以还需要考虑编辑距离是 2 的情况。99%的正确词和错误词的编辑距离 2 以内。因此，对于拼写检查来说，查找出编辑距离在 2 以内的候选正确词 e 的集合即可。这是一个模糊匹配的问题。

6.3.1 模糊匹配问题

语料库文本中可能会存在拼写错误。例如，"banana"可能会被错误地写成字面上相似的"bannana"。测量两个字符串之间相似性的标准方法是编辑距离。编辑距离是插入、删除和替换字母的数量。"bannana"和"banana"之间的编辑距离是 1，因为插入了 1 个"n"。

当搜索查询没有给出结果时，可以搜索编辑距离为 1 的所有查询。因此，如果遇见单词"bannana"，可以生成编辑距离为 1 的所有单词，并从正确词表中搜索那些单词。如果仍然没有产生任何结果，可以生成编辑距离为 2 的所有单词，以此类推。问题是搜索词的数量是巨大的，可以插入、删除和替换任何字母，所以仅供第一个字母使用的单词可能是"aanana""canana""danana"等。如果使用完整的 Unicode 字符编码表，这尤其不可能。

假设正确词表组成如图 6-1 所示的字典树（trie tree，又称为单词查找树）：

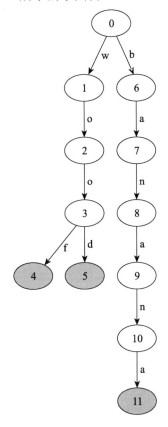

图 6-1 包含 "woof" "wood" "banana" 的字典树

当在"bannana"的编辑距离 1 内搜索这棵字典树中的单词时，整个左分支是无关紧要的。可以从根节点开始搜索，然后通过"w"向左移动。继续搜索，因为当走得更远时，可能会找到"wannana"。继续向下找到"o"，所以现在在节点 2 处使用前缀"wo"。可以停止搜索，因为没有以"wo"开头的单词可以在编辑距离为 1 的"bannana"内。所以回溯并继续沿着"b"开始向右分支。这次一直到节点 11 并找到"banana"，这是一个匹配。可以从搜索中删除一些子树，而不是搜索整棵字典树。在真实的索引中，可能有数百万词。"wo"下的子树可能很大，不需要搜索任何子树。

这就是 Levenshtein 自动化的用武之地。给定一个单词和最大编辑距离，可以构建一个 Levenshtein 自动机。用户一个接一个地给这个自动机提供字母，然后自动机可以告诉用户是否需要继续搜索。在这个例子中，为"bannana"创建了一个 Levenshtein 自动机，$N=1$。从根节点开始并将其输入"w"。那么可以匹配吗？自动机说是的。继续下去并喂它"o"。那么可以匹配吗？自动机说不，所以回溯到根，也回溯了自动机状态！喂自动机"b"并询问它是否仍能匹配？它说是的。继续沿着正确的路径喂它，直到到达节点 11。然后问"这匹配吗？"自动机说是的，找到了搜索结果。这是 Python 中 Levenshtein 自动机的接口：

```python
class LevenshteinAutomaton:
    def __init__(self, string, n):
        ...

    def start(self):
        #返回开始状态
        ...

    def step(self, state, c):
        #给定状态和字符时返回下一个状态
        ...

    def is_match(self, state):
        ...

    def can_match(self, state):
        ...
```

有一个构造函数，它接受查询字符串和最大编辑距离 n。有一个返回启动状态的 start() 方法和一个返回给定状态和字符的下一个状态的 step(state, c) 方法，还有可告诉状态是否匹配的 is_match(state) 方法，并且有告诉如果输入更多字符，状态是否可以匹配的 can_match(state) 方法。在 can_match() 方法中剪枝搜索树。如下是一个例子：

```
automaton = LevenshteinAutomaton("bannana", 1)

state0 = automaton.start()
state1 = automaton.step(state0, 'w')
automaton.can_match(state1) #返回 True,"w"可以在距离 1 的条件下匹配"bannana"
state2 = automaton.step(state1, 'o')
automaton.can_match(state2) #返回 False, "wo"不能在距离 1 的条件下匹配"bannana"
```

接下来根据编辑距离算法来实现 Levenshtein 自动机的接口。

考虑如何衡量两个字符串 S 和 T 之间的差异 distance(S, T)。例如，gold 和 good 之间只差一个字母，而 gold 和 bad 之间差更多的字母。所以，distance(gold, good) < distance(gold, bad)。

给定两个字符串：$S = s_1 s_2 \ldots s_m$ 和 $T = t_1 t_2 \ldots t_n$

假设一个打字员，把单词 T 错误地输入成为单词 S 了，这个打字员需要若干次操作来改正这个错误。对于把 S 转换到 T 所需要的一系列编辑操作，想找到最小的花费 D(S,T)。编辑操作包括下面 3 种：
- 使用 T 中的一个字符替换 S 中的一个字符。
- 删除 S 中的一个字符。
- 插入 T 中的一个字符。

把 D(S,T)叫作编辑距离。D(S,T)是一个不小于零的整数。例如，把 gold 转换成 good 最少需要 1 次编辑操作，也就是把 l 替换成 o，所以 D(gold, good)=1。而把 gold 转换成 bad 最少需要 3 次编辑操作，所以 D(gold, bad)=3。因此，D(gold, good)< D(gold, bad)。

distance(gold, good)可以被看成是这样计算得到的：将 gold 的最后一个字符 d 替换成 good 中的 d。因为这两个字符相同，所以 d 这个字符并没有导致编辑距离增加，代价是 0。

对于 S 中的字符 s 和 T 中的字符 t 来说，假设 t 是由字符 s 替换得到的。因为 t 的存在，导致编辑距离增加的值叫作代价 c。根据三角形中两边长度之和大于第三条边，对于代价 c，假设三角不等式：

$$c(a,c) \leqslant c(a,b) + c(b,c)$$

从 a 转换成 c 的代价不大于要通过 b 这个步骤的代价。c(a,c)约减成两个字符之间的变换。也就是说，每个字符最多改变一次。类似的，日常生活中往往不通过中介找房东直接租房成本低。

如果 s 和 t 相同，则代价是 0。如果 s 和 t 不同，则代价是 1。代价模型用伪代码表示为：

$$if(s!=t)\ c(s,t)=1$$

$$if(s==t)\ c(s,t)=0$$

distance(goods, good)可以被看成是这样计算得到的：源串删除最后一个源字符 s 就得到了目标字符串 good。

distance(goo, good)可以被看成是这样计算得到的：源串插入最后一个目标字符 d 就得到了目标字符串。

假设把 S 的子串 $s_1...s_i$ 写作 S_i，而把 T 的子串 $t_1...t_j$ 写作 T_j。$D_{i,j}= D(S_i,T_j)$

把 S 转换到 T 有 3 种结束的可能：
- 最后一个字符是从源字符替换成目标字符。如果源字符串和目标字符串的最后一个字符正好相等，则代价是 0；否则，代价是 1。用 t_n 替换 s_m：$D_{m,n} = D_{m-1,n-1} + c(s_m, t_n)$。
- 删除最后一个源字符就得到了目标字符串。删除 s_m：$D_{m,n}= D_{m-1,n} + 1$。
- 插入最后一个目标字符就得到了目标字符串。插入 t_n：$D_{m,n} = D_{m,n-1} + 1$。

$D_{i,j}$ 由 $D_{i-1,j-1}$、$D_{i-1,j}$、$D_{i,j-1}$ 推出。如图 6-2 所示计算 $D_{i,j}$。

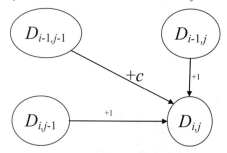

图 6-2　计算编辑距离问题分解图

把 levenshtein 转换成 meilenstein 的过程如下：

图 6-3 中，=表示匹配上了；o 表示替换；+表示插入；-表示删除。

l	e	v	e	n	s	h	t	e	i	n
o	=	+	o	=	=	-	=	=	=	=
m	e	i	l	e	n	s	t	e	i	n

图 6-3　meilenstein 和 levenshtein 的编辑距离

形式化的写法：

$D_{m-1,n}$ -> $D_{m,n}$ 就是 -s

$D_{m-1,n-1}$ -> $D_{m,n}$ 就是 s o t

$D_{m,n-1}$ -> $D_{m,n}$ 就是 +t

如果 D(gol, goo)已知，则 D(gold, good)也就知道了，因为 D(gold, good)和 D(gol, goo)的值相同。所以，可以认为 D(gold, good)依赖于 D(gol, goo)的值。而 D(gol, goo)又依赖于 D(go, go)的值，D(gol, goo)比 D(go, go)的值大 1。

一个问题可以由子问题推出。图 6-4 是计算 D(abc,abb)的问题分解图。其中，D(abc,abb)依赖于 D(abc,ab)、D(ab,abb)和 D(ab,ab)的结果。D(abc,ab) -> D(abc,abb)就是目标单词 T 增加一个字符 b。D(ab,abb) -> D(abc,abb)就是源单词 S 删除一个字符 c。D(ab,ab) -> D(abc,abb)就是源单词 S 最后一个字符 c 替换成 b。这里的 D(ab,ab)在图 6-4 中重复出现。也就是说，分解计算 D(abc,abb)时，会重复计算 D(ab,ab)。

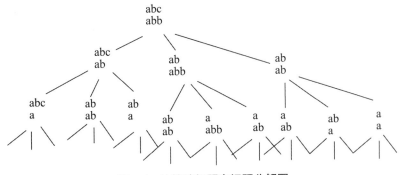

图 6-4 计算编辑距离问题分解图

编辑距离就是从源串转换成目标串的最小代价所以取最小值,走近路。就像出租车在一般情况下不能绕远路;否则打车费可能高了。

求解编辑距离的循环等式(recurrence equation):

$$D_{m,n} = \min \begin{cases} D_{m-1,n-1} & + & c(s_m, t_n) \\ D_{m-1,n} & + & 1 \\ D_{m,n-1} & + & 1 \end{cases} \qquad (6\text{-}3)$$

需要计算所有的 $D_{i,j}$ 当 $0 \leqslant i \leqslant m, 0 \leqslant j \leqslant n$。

分治法的实现方法采用自顶向下递归求解,代码如下:

```
def LD(s, t):
    if s == "": #如果源字符串长度是0,则编辑距离就是目标字符串的长度
        return len(t)
    if t == "": #如果目标字符串长度是0,则编辑距离就是源字符串的长度
        return len(s)
    #循环等式
    if s[-1] == t[-1]:
        cost = 0
    else:
        cost = 1
    #把当前问题分解成三个子问题
    res = min([LD(s[:-1], t)+1,           #源串删除一个字符成为目标串
               LD(s, t[:-1])+1,            #源串插入一个字符成为目标串
               LD(s[:-1], t[:-1]) + cost]) #源串替换一个字符成为目标串
    return res
```

如果采用分治法,则有些子问题的计算是重复的。所以要把子问题的计算结果存起来。动态规划方法采用从底向上计算。通用的方法是用散列表存储计算的中间结果,这里采用二维数组。实现代码如下:

```
def iterative_levenshtein(s, t):
    """
        iterative_levenshtein(s, t) -> ldist
```

```python
    ldist 是字符串 s 和 t 之间的 Levenshtein 距离
    """
    rows = len(s)+1
    cols = len(t)+1
    dist = [[0 for x in range(cols)] for x in range(rows)] #最小花费矩阵
    #源串前缀可以通过删除转换为空字符串
    for i in range(1, rows):
        dist[i][0] = i
    #可以通过插入字符从空源串创建目标串前缀
    for i in range(1, cols):
        dist[0][i] = i

    for col in range(1, cols):
        for row in range(1, rows):
            if s[row-1] == t[col-1]:
                cost = 0
            else:
                cost = 1
            dist[row][col] = min(dist[row-1][col] + 1,      #删除
                                 dist[row][col-1] + 1,      #插入
                                 dist[row-1][col-1] + cost) #替换
    for r in range(rows):
        print(dist[r])

    #在迭代步骤结束后,距离在单元 d[row, col].
    return dist[row][col]
```

如果逐行填充矩阵,那么就不需要存储完整的矩阵。一旦计算了第 i 行,就可以扔掉第 i-1 行,所以只需要存储一行。这可用于实现 Levenshtein 自动机:作为其状态,使用 Levenshtein 矩阵的单行。给定一个状态(矩阵中的一行)和一个字符 c,可以用上面的公式计算下一个状态(矩阵中的下一行)。可以通过查看行中的最后一个条目来实现 is_match,该条目包含查询与输入到自动机的字母串之间的编辑距离。如果该数字小于 n,就匹配了。由于行中的数字永远不会因步进而减少,因此实现 can_match 方法也很简单:只需检查行中的任何数字是否小于 n。如果有这样的数字,那么,如果输入与查询字符串匹配的字母,仍然可以匹配,因此 cost = 0。LevenshteinAutomaton 类的实现代码如下:

```python
class LevenshteinAutomaton:
    def __init__(self, string, n):
        self.string = string
        self.max_edits = n

    def start(self):
        return range(len(self.string)+1)

    def step(self, state, c):
```

```
            new_state = [state[0]+1]
            for i in range(len(state)-1):
                cost = 0 if self.string[i] == c else 1
                new_state.append(min(new_state[i]+1, state[i]+cost, state[i+1]+1))
            return new_state

    def is_match(self, state):
        return state[-1] <= self.max_edits

    def can_match(self, state):
        return min(state) <= self.max_edits
```

可以直接使用基于 step() 的自动机搜索索引, 或者从 Levenshtein 自动机中构建一个 DFA（确定有限状态机）。DFA 具有有限数量的状态, 但自动机具有无限数量的状态, 每次步进中的数字都会增加。问题在于, 许多状态都是等同的。例如, 如果 $n=2$, 那么这些状态都是等价的:

```
[3, 2, 1, 2, 3, 4, 5]
[500, 2, 1, 2, 500, 500, 500]
[3, 2, 1, 2, 3, 3, 3]
```

一旦数字超过 n, 无论是 3、5 还是 500 都无关紧要。它永远不会导致匹配。因此, 不是无限期地增加这些数字, 也可以将它们保持在 3 上。在代码中改变如下这一行

```
    return new_state
```

成为：

```
    return [min(x,self.max_edits+1) for x in new_state]
```

现在状态的数量是有限的！可以递归地探索自动机状态, 同时跟踪已经看过的状态而不再访问它们。然而, 还有一件事需要：自动机需要告诉用户从给定状态尝试哪些字母。只需要尝试实际出现在查询字符串中相关位置的字母。如果查询字符串是 "banana", 则不需要分别尝试 "x" "y" 和 "z", 因为它们都具有相同的结果。此外, 如果状态中的条目已经是 3, 则不需要尝试字符串中的相应字母, 因为它无论如何都不会导致匹配。这是计算要尝试的字母的代码：

```
def transitions(self, state):
    return set(c for (i,c) in enumerate(self.string) if state[i] <= self.max_edits)
```

现在可以递归地枚举所有状态并将它们存储在 DFA 转换表中：

```
counter = [0]
states = {}
transitions = []
matching = []
```

```
lev = LevenshteinAutomaton("woof", 1)

def explore(state):
    key = tuple(state) #列表不能在Python中进行哈希处理,因此把列表转换为元组
    if key in states: return states[key]
    i = counter[0]
    counter[0] += 1
    states[key] = i
    if lev.is_match(state): matching.append(i)
    for c in lev.transitions(state) | set(['*']):
        newstate = lev.step(state, c)
        j = explore(newstate)
        transitions.append((i, j, c))
    return i

explore(lev.start())
```

图 6-5 是 "woof" 和 $n = 1$ 对应的 DFA。

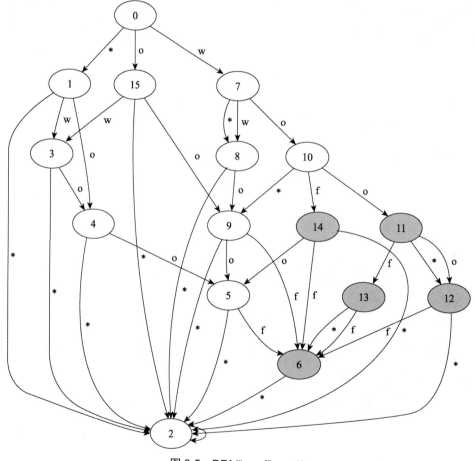

图 6-5　DFA("woof", $n = 1$)

深色节点是可接受状态。

如果为大小为 1000 的查询字符串构建一个 $n = 2$ 的自动机，那么每个状态都是一个大小为 1001 的矩阵行，但它最多包含 5 个不等于 3 的条目！为什么还在计算其他 996 个等于 3 的条目？这种问题的标准解决方案是使用稀疏向量。用户只存储不等于 3 的那些，而不是存储所有条目。不是这样：

```
state = [3,3,3,3,3,3,3,3,3,3,3,2,1,2,3,3,3,3,3,3]
```

存储两个小数组：

```
values = [2,1,2]
indices = [11,12,13]
```

使用此表示法计算 step（state,c）会变得有点困难：

```python
class SparseLevenshteinAutomaton:
    def __init__(self, string, n):
        self.string = string
        self.max_edits = n

    def start(self):
        return (range(self.max_edits+1), range(self.max_edits+1))

    def step(self, (indices, values), c):
        if indices and indices[0] == 0 and values[0] < self.max_edits:
            new_indices = [0]
            new_values = [values[0] + 1]
        else:
            new_indices = []
            new_values = []

        for j,i in enumerate(indices):
            if i == len(self.string): break
            cost = 0 if self.string[i] == c else 1
            val = values[j] + cost
            if new_indices and new_indices[-1] == i:
                val = min(val, new_values[-1] + 1)
            if j+1 < len(indices) and indices[j+1] == i+1:
                val = min(val, values[j+1] + 1)
            if val <= self.max_edits:
                new_indices.append(i+1)
                new_values.append(val)

        return (new_indices, new_values)

    def is_match(self, (indices, values)):
        return bool(indices) and indices[-1] == len(self.string)
```

```
    def can_match(self, (indices, values)):
        return bool(indices)

    def transitions(self, (indices, values)):
        return set(self.string[i] for i in indices if i < len(self.string))
```

6.3.2 英文拼写检查

使用 TextBlob 的 correct 函数体验拼写纠错。

```
>>> blob = TextBlob('Analytics Vidhya is a gret platfrm to learn data scence')
>>> blob.correct()
TextBlob("Analytics Vidhya is a great platform to learn data science")
```

使用拼写检查功能可以检查建议的单词列表及其置信度。

```
blob.words[4].spellcheck()
>> [('great', 0.5351351351351351),
 ('get', 0.3162162162162162),
 ('grew', 0.11216216216216217),
 ('grey', 0.026351351351351353),
 ('greet', 0.006081081081081081),
 ('fret', 0.002702702702702703),
 ('grit', 0.0006756756756756757),
 ('cret', 0.0006756756756756757)]
```

先看一下英文拼写出错的可能性有多大。对英文报关公司名的 101919 条统计，有拼写出错的为 16663 条，出错概率为 16.35%。因为大部分是正确的，所以若所有的词都在正确词表中，则不必再查找错误。

正确词的词典格式每行一个词，分别是词本身和词频。样例如下：

```
biogeochemistry : 1
repairer : 3
wastefulness : 3
battier : 2
awl : 3
preadapts : 1
surprisingly : 3
stuffiest : 3
```

根据发音可以计算用户输入词和正确词表的相似度，也可以根据字面的相似度来判断是否输入错误，并给出正确的单词提示，还可以参考一下开源的拼写检查器（spell checker）的实现，例如 Aspell（http://aspell.net/）或者 LanguageTool（http://www.languagetool.org/）。

特别的，考虑公司名称中的拼写错误。一般首字母拼写错误的可能性很小。可以简单地先对名称排序，然后再比较前后两个公司名称就可以检测出一些非常相似的公司名称。

6.3.3 中文拼写检查

对于汉语的查询纠错,系统预测错误的并不是要产生一个有语法错误的纠错串,而是与原串相比,语义的偏离!怎样才能通过一个带有错误的查询串猜测到用户的真实查询意图,给出一个满足条件的纠错串,这个纠错串代表了用户的查询意图,这个是查询纠错的目的,而语义的偏离才是影响系统性能的首要因素。

和英文拼写检查不一样,中文用户输入的搜索词串的长度更短,从错误的词猜测可能的正确输入更加困难。这时候需要更多地借助正误词表,词典文本格式如下:

```
代款:贷款
阿地达是:阿迪达斯
诺基压:诺基亚
飞利蒲:飞利浦
寂么沙洲冷:寂寞沙洲冷
欧米加:欧米茄
欧米枷:欧米茄
爱力信:爱立信
西铁成:西铁城
瑞新:瑞星
登心绒:灯心绒
```

这里,前面一个词是错误的词条,后面是对应错误词条的正确词条。为了方便维护,还可以把这个词库存放在数据库中:

```
CREATE TABLE CommonMisspellings (
        [misword] [varchar] (50) COLLATE Chinese_PRC_CI_AS NULL , --错误词
    [rightword] [varchar] (50) COLLATE Chinese_PRC_CI_AS NULL  --正确词
)
```

除了人工整理,还可以从搜索日志中挖掘相似字串来找出一些可能的正误词对。比较常用的方法是采用编辑距离来衡量两个字符串是否相似。编辑距离就是用来计算从原串(s)转换到目标串(t)所需要的最少的插入、删除和替换的数目。例如,源串是"诺基压",目标串是"诺基亚",则编辑距离是1。

当一个用户输入错误的查询词没有结果返回时,他可能会知道输入错误,然后用正确的词再次搜索。从日志中能找出来这样的行为,进而找出正确/错误词对。

例如,日志中有这样的记录:

```
2007-05-24 00:41:41.0781|DEBUG |221.221.167.147||瑜伽服|2
…
2007-05-24 00:43:45.7031|DEBUG|221.221.167.147||瑜伽服|0
…
```

可以挖掘出如下一些错误/正确词对:

```
谕伽服:瑜伽服
```

```
落丽塔:洛丽塔
巴甫洛:巴甫洛夫
hello kiitty:hello kitty
…
```

除了根据搜索日志挖掘正误词表,还可以根据拼音或字形来挖掘。例如,根据拼音挖掘出"周杰论:周杰伦";根据字形挖掘出"浙江移动:浙江移动"。

第 7 章
Web 搜索案例分析

本章介绍使用 Elasticsearch 实现的医药垂直搜索引擎和电商站内搜索。

7.1 医药垂直搜索引擎

利用 Python 语言程序读取 Excel 表格中的网址定向采集医生信息。使用 openpyxl 模块实现 Excel 文件的读取。

```
from openpyxl import load_workbook

#load excel file
workbook = load_workbook(filename="医生信息与出诊时间.xlsx")

#open workbook
sheet = workbook["输出-医生信息"]

url = sheet["G2"]
print("G2 cell value:",url.value)
```

用户通过界面选择 Excel 文件:

```
from pathlib import Path
import PySimpleGUI as sg

sg.theme("DarkBlue")

layout = [
    [sg.Text('Doctor list file'), sg.InputText(key='-file1-'), sg.FileBrowse()],
    [sg.Button("Go")],
]

window = sg.Window('Crawler', layout)
```

```
while True:
    event, values = window.read()
    if event == sg.WINDOW_CLOSED:
        break
    elif event == "Go":
        filename = values['-file1-']
        while True:
            if not Path(filename).is_file():
                if filename == '':
                    sg.popup_ok('Select a file to go !')
                else:
                    sg.popup_ok('File not exist !')
                filename = sg.popup_get_file("", no_window=True)
                if filename == '':
                    break
                window['-file1-'].update(filename)
            else:
                print('File is ready !')
                break
window.close()
```

使用 pyppeteer 库采集医生信息:

```
import asyncio
from pyppeteer import launch

async def main():
    browser = await launch()
    page = await browser.newPage()
    await page.goto('https://www.haodf.com/doctor/2685.html')

    element = await page.querySelector('.brief-content-wrap')
    brief_str = await (await element.getProperty('textContent')).jsonValue()
    print(brief_str)
    await browser.close()

asyncio.get_event_loop().run_until_complete(main())
```

将医生信息写入索引库:

```
es = Elasticsearch()

e1={
"url":docURL,
"introduction": brief_str
}
```

```
res = es.index(index='doctors',id=1,body=e1)
print(res)
```

查询医生信息。

```
res=
es.search(index='doctors',body={'query':{'match':{'introduction':queryStr}}})
print("match introduction Got %d Hits:" % res['hits']['total']['value'])
print(res['hits']['hits'])
```

7.2 内容管理系统搜索

CastleCMS 为要求最严苛的组织提供网站、内联网和 Web 应用程序的支持。CastleCMS 建立在可靠的开源内容管理系统（CMS）Plone 之上，提供了世界一流的内容管理功能和最大安全性的独特组合。

除了 Plone 标准功能外，CastleCMS 还包括：

- 登录/退出登录支持。
- 内容归档到 Amazon S3 存储。
- 大文件自动移至 S3 存储。
- Redis 缓存支持。
- 高级内容布局编辑器。
- 改进的管理工具栏。
- 直观的内容创建和组织。
- Elasticsearch 集成。
- 根据社交媒体影响调整的搜索结果。
- Celery 任务队列集成（异步操作）：
  - PDF 生成；
  - 视频转换；
  - Amazon S3 交互；
  - 大批量项目的复制和粘贴；
  - 大批量项目的删除和重命名。
- 高级内容贴图：
  - 地图；
  - 视频；
  - 音频；
  - 滑块；
  - 画廊；

- ➢ 目录。
- 音频和视频内容。
- 自动将视频转换为 Web 兼容格式。
- 使用 Google Analytics API 基于流行度的搜索权重。
- 内容别名管理。
- Disqus 评论集成。
- reCAPTCHA 验证码集成。
- FullCalendar 集成。
- Google 业务元数据。
- 带有可选短信支持的紧急通知系统。
- 在各种设备尺寸上预览内容。
- 地图内容。
- KML 信息流。
- 社交媒体与 Twitter、Facebook、Pinterest 的整合。
- Etherpad 协作空间支持。
- 从文件中剥离元数据。
- 能够以其他用户的身份查看站点。
- 审计日志、用户活动报告。
- 会话管理、检查和终止。
- 分析仪表板。
- 对上传的图像和文件进行重复数据删除。
- 垃圾桶/回收站。
- 双因素身份验证。

在 macOS 上的开发设置如下：

（1）brew install redis elasticsearch libav python。

（2）git clone git@github.com:castlecms/castle.cms.git。

（3）cd castle.cms。

（4）virtualenv -p python2.7。

（5）bin/pip install --upgrade pip。

（6）bin/pip install -r requirements.txt。

（7）bin/buildout。

（8）在单独的终端窗口中运行 elasticsearch, redis-server, bin/instance fg。

（9）浏览到 http://localhost:8080/。

（10）在开发实例中创建站点后，运行 init-dev 脚本以填充模板：bin/instance run castle/cms/_scripts/init-dev.py。